Darwin's Falling Sparrow

Victorian Evolutionists and the Meaning of Suffering

Kristin Johnson

Prometheus Books

Essex, Connecticut

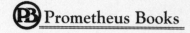

An imprint of Globe Pequot, the trade division of The Rowman & Littlefield Publishing Group, Inc.
4501 Forbes Blvd., Ste. 200
Lanham, MD 20706
www.rowman.com

Distributed by NATIONAL BOOK NETWORK

British Library Cataloguing in Publication Information Available

Library of Congress Cataloging-in-Publication Data
Names: Johnson, Kristin, 1973– author.
Title: Darwin's falling sparrow : Victorian evolutionists and the meaning of suffering / Kristin R. Johnson.
Description: Lanham : Prometheus, [2024] | Includes bibliographical references and index. | Summary: "The book applies a biographical, narrative lens to explore what people in the past believed and why, and how and why those beliefs—about God, nature, history, and human agency—changed over time"— Provided by publisher.
Identifiers: LCCN 2023012804 (print) | LCCN 2023012805 (ebook) | ISBN 9781633888746 (cloth) | ISBN 9781633888753 (epub)
Subjects: LCSH: Evolution (Biology)—Philosophy. | Darwin, Charles, 1809–1882—Influence. | Evolution (Biology)—Religious aspects—Anglican Communion. | Suffering—Religious aspects—Anglican Communion. | Philosophy, English—19th century. | Great Britain—Religion—19th century.
Classification: LCC QH360.5 .J64 2024 (print) | LCC QH360.5 (ebook) | DDC 576.8/2--dc23/eng/20230622
LC record available at https://lccn.loc.gov/2023012804
LC ebook record available at https://lccn.loc.gov/2023012805

∞™ The paper used in this publication meets the minimum requirements of American National Standard for Information Sciences—Permanence of Paper for Printed Library Materials, ANSI/NISO Z39.48-1992

CONTENTS

INTRODUCTION

Are not two sparrows sold for a farthing?
and one of them shall not fall on the ground without your Father.
But the very hairs of your head are all numbered.
Fear ye not therefore, ye are of more value than many sparrows.
                                        —*MATTHEW 10:29–31*

ON MARCH 26, 1843, A GRIEVING FATHER NAMED WILLIAM OWEN answered a letter of condolence from the well-known British naturalist Charles Darwin. Owen's beloved son, a soldier, had died in India. Within the space of a few weeks, yellow fever and cholera had "carried off the greatest Part of his fine Regiment without their having fired a shot." Darwin's letter does not survive, but we can tell from Owen's reply that Darwin hoped Owen could find some comfort in the fact that his son had left this world without having suffered the misfortunes and disappointments of most men. For, as Owen thanked Darwin for his sympathetic words, he wrote that he did indeed find comfort in the fact his son had lived a happy life. Owen then added that he also found solace in the belief that "all is for the best & that the blow has been struck in mercy by that Almighty Spirit who we are bound to believe cannot err, & without whose knowledge & will not a Sparrow falls."[1]

Charles could well sympathize with Owen's loss. He and his wife Emma had lost an infant son five months before. (They would lose two more children within a decade.) But by the time he wrote this letter of condolence, Darwin suspected that God simply did not govern the world in the way Owen believed. Darwin was drifting away from belief in the

God without whose "knowledge & will not a Sparrow falls." By the time of their compassionate exchange, these fathers lived, thought, and grieved on opposite sides of a growing chasm created by developments in British science, religion, and even politics. (As we will see, these things were tightly linked together.) Owen trusted that a divine purpose and meaning existed in his son's death. Darwin had his doubts. Owen appealed to the faith that all affliction is under the attentive governance of a close, personal, and loving God. Darwin was slowly dispensing with this belief under the strong suspicion that, despite the comforting passage in the Book of Matthew, God was not paying any attention when sparrows fell to the ground, infants died, or sons did not come home.

This book is about how attempts to explain the origin and meaning of suffering influenced the development of and response to Darwin's ideas. Since the eighteenth century, explanations of why God allows so much evil and suffering in the world have been called "theodicies." The word theodicy—from *théo* (God) and *dicée* (justice, to form)—was coined by the German philosopher Gottfried Leibniz in 1710, but of course attempts to explain why the all-powerful, all-good, and all-knowing God of most monotheist traditions permits evil and suffering extend much further back in time. Within the dominant Christian traditions of Darwin's Britain, various theodicies prevailed: some held that God must have unknown yet good reasons for inflicting or permitting suffering; others emphasized that a creation in which free will exists is a great good, but that free will comes with a cost, namely, that humans may act wrongly and in doing so produce evil and suffering for themselves or others. But most believed that all suffering was, ultimately, heaven-sent.

Given Victorian Christians' belief that God is "of infinite power, wisdom, and goodness," the message of the New Testament's "falling sparrow" passage was supposed to be clear: human beings, of "more value than many sparrows," were under God's watchful care. Amid tremendous diversity, most Christians agreed: suffering came directly from God, and therefore a divine purpose existed in every affliction, whether fallible "man" could discern that purpose or not. In this world, the appropriate response to affliction was widely agreed upon, even if it needed constant repeating: trust in God's will and the possibility of reunion with loved ones in heaven. This response did not necessarily equate to inaction in the face of suffering (after all, like most religious traditions, Christians supported both charity and medical care), but belief in God's providential control over human lives provided meaning and purpose as to why such suffering existed in the first place.

We know from mortality tables, diaries, and personal correspondence that for Darwin and his fellow Victorians, the "falling sparrow" at stake in attempts to understand the suffering in the world was often a beloved child. Throughout the nineteenth century, tuberculosis, infant diarrhea, dysentery, typhoid fever, scarlet fever, whooping cough, diphtheria, pneumonia, measles, smallpox, and other ailments took children away from their parents. "Let us remember how common the Calamity is," warned a minister named Reverend Doddridge, "few Parents are exempt from it; some of the most pious and excellent have lost amiable Children." When Darwin was born in 1809 the mortality rate for children under the age of five in the United Kingdom was 315 per 1,000 births. When Darwin died in 1882 the rate had dropped to about 220—still unimaginably high to most people today. (Today, the rate hovers at 4 per 1,000 births.)[2]

Almost every individual involved in the story of Darwin's life and work lost at least one child. (Most of those who did not, including Charles Lyell, Harriet Martineau, and Asa Gray, did not have any children to lose.) Sometimes they lost several children. William Paley (from whom Darwin learned the argument that species are created by a benevolent God) and his wife Jane lost two infants, John and Francis. Charles Darwin's evolutionist-grandfather, Erasmus Darwin, buried three of his twelve legitimate, and therefore known, children when they were infants. Robert Chambers, who published a book on evolution in 1844, lost a daughter to scarlet fever and three other children to unknown ailments. One of Chambers's most ardent critics, Hugh Miller, lost a daughter to smallpox. Three of Charles and Emma Darwin's children died. The co-discoverer of natural selection, Alfred Russel Wallace, lost a son, Herbert, to scarlet fever when the boy was six. Darwin's ardent defender Thomas Henry Huxley lost his firstborn son, four-year-old Noel, to scarlet fever. The botanist Joseph Dalton Hooker, to whom Darwin first confessed his theory of evolution, lost his little girl Maria when she was six. For each illness that ended in loss, these parents sat at their children's bedsides amid fevers and coughing fits, not knowing whether their children would live or die. The atheist Annie Besant, who took Darwin's theories into realms at which he balked (including a campaign for contraception), almost lost her seven-month-old daughter to whooping cough. One could go on and on: Darwin's defenders, opponents, and everyone in between experienced the very real threat or the reality of child loss.

Because developments in medicine, public health movements, and higher living standards have protected many from something that was once so common, it can be difficult to imagine a world in which even the children of the most privileged and prosperous could be snatched away within the space of a few days after fever entered a home. But once we set aside the comforting myth that parents in eras of high child mortality did not suffer as much because they avoided becoming attached to their children, these losses highlight the potential stakes involved in Victorians' confrontations with alternative explanations of the world. (The claim that parents in previous times did not become attached to their children was apparently first told by social historian Philip Aries in his 1960 book *Centuries of Childhood*. In fact, historians have provided extensive evidence that parents have both loved and grieved for their children throughout time.)[3] We can learn important things about this past by attending to how Darwin, his allies, and his critics debated "what the world is like and why" within a realm of human experience in which the stakes in deciding were very high indeed.

As a young man, Charles Darwin agreed with William Owen's faith that great affliction, including even a beloved son's death, were blows "struck in mercy by that Almighty Spirit who we are bound to believe cannot err, & without whose knowledge & will not a Sparrow falls." We know this because he described himself as holding quite orthodox beliefs when he set off on the voyage of the HMS *Beagle* in December 1831. Indeed, at the time he was a student at Cambridge University with plans to become a minister in the Church of England, a career plan that would have required him to assent to the Thirty-Nine Articles of the Anglican Faith (codified in 1562 when England split from the Catholic Church and became Protestant). Those articles included the belief that Christ had vicariously atoned for humanity's sins, that Christ waited in heaven "to judge all Men at the last day," and that God had given human beings binding moral commandments and promised rewards and punishments in the hereafter (everlasting life in heaven or eternal damnation in hell).

Anglican Christianity emphasized God's close personal governance over creation (even a sparrow "shall not fall on the ground without your Father"). It also upheld humanity as the one species created in God's image ("Fear ye not therefore, ye are of more value than many sparrows"). The first article described God as of "infinite power, wisdom, and goodness; the Maker, and Preserver of all things both visible and invisible."[4] These attributes—infinite power, wisdom, goodness, and being the "maker and preserver of all things"—will be central to our story. The article's statement that God makes and preserves all things is based on the Old Testament's account of the history of creation in Genesis 1 and 2. There one also finds an explanation for the entrance of suffering into God's creation. Created in the Garden of Eden, Adam and Eve had disobeyed God's commandment not to eat from the Tree of Knowledge. Their punishment extended vicariously to all humanity. This story had made sense of the existence of suffering for more than a thousand years— from the enmity between husbands and wives; to the pain of childbirth; to humanity's fate of labor, disease, and physical death. Adam and Eve's commission of the first "original" sin also explained humanity's extraordinary capacity for both good and evil (including the fact humans know what is right yet often choose wrongly and therefore sin), and thus why a messiah was needed to redeem humanity for its sins. For Christians, that messiah was Jesus Christ, who, according to the New Testament, descended as the son of God to atone, by His sacrifice, for humanity's rebellion and provide a path to eternal salvation for those who believed in Him.

What does any of this have to do with debates over the origin of species? To answer this question, we must immerse ourselves in an era of British history during which Christianity and the study of the natural world were tightly intertwined. When Darwin was born, most European naturalists (those who study the natural history, behavior, and comparative anatomy of animals and plants) saw themselves as crucial allies in the campaign to uphold Christianity's explanation of the origin and meaning of suffering, including child loss. They argued, for example, that the purposeful parts of animals and plants provided rational, scientific demonstrations of Scripture's message of God's close personal governance of the world. In other words, the study of nature vindicated the Book of Matthew's promise that not a sparrow fell to the ground without God's attentive and benevolent will. Furthermore, the British powers-that-be were intent on spreading that argument throughout a growing empire.

By the time Darwin died in 1882, the options available for explaining the world, from the origin and fate of sparrows to child loss, had expanded a great deal. The history of these options will not be a simple tale. For example, surprisingly few readers of Darwin's *On the Origin of Species* came to the conclusion that his theory required a repudiation of belief in a benevolent God. Darwin's most committed defender in the United States, the Harvard botanist Asa Gray, believed that a Darwinian "struggle for existence" was precisely the kind of pervasive suffering one should expect from a correct interpretation of the Bible. After all, Scripture described the creation as fully partaking in the fall of man (for example, Romans 8:22 states, "For we know that the whole creation groaneth and travaileth in pain together until now"). Indeed, Gray famously argued that Darwin had turned suffering, the great challenge of belief in an all-good, all-powerful, and all-knowing God, to "creative account" by explaining the apparent waste and suffering as part of a divine, progressive, "economical process." As historians have shown, some of Darwin's most committed allies in revising explanations of divine governance in the world were evangelical Christians. Of course, some readers were not so sure that Darwin's theory could be reconciled with belief in the fatherly God who was present at each sparrow's fall. For those intent on breaking the power of Anglican Christianity in Britain and its Empire, that difficulty proved a major part of the theory's appeal. This is all to say that this will be a complicated story, not a simple one.[5]

Attending to the problem of child loss in the development and reception of Darwin's work reminds us that debates over the idea of evolution are about much more than how to explain the shape of finch beaks. Questions about God's existence and governance, the origin and meaning of suffering, the possibilities of progress, and the power of human agency were all at stake. Not surprisingly, the broader questions with which Victorians wrestled continue to influence stances on a range of pressing issues, from climate change to genetic engineering. Individuals continue to differ on how to explain the origin and meaning of suffering, the possibilities and best means of progress, and the power and limits of human action. These differences in beliefs, values, and assumptions generally exist "between the lines" of debates that concern science, yet noticing them is crucial to understanding why stances on science seem to differ so much. History is one of our most powerful tools for learning how to notice and map the factors at play in debates that concern science and the amelioration of suffering in the present day.

That said, studying the assumptions, beliefs, and values of the past poses a number of challenges. We tend to have strong beliefs regarding God, science, nature (and evolution in particular), and the origin of and best response to suffering. Yet doing good history and accurately understanding stances (especially those that differ from our own) requires that we try to suspend our own beliefs and assumptions about what is true or false, right or wrong. Failure to do so can result in misdiagnoses of the complex origins and meanings of particular ideas in the past, whether we believe those ideas led to a more accurate portrait of the world or not, and whether we believe those ideas led to liberation or oppression. And of course, misunderstanding the past ultimately leads to inaccurate maps of the present.

As historians David Lindberg and Ronald Numbers have eloquently warned, studying this past through our own "battle-scarred glasses" (from either side of today's debates over evolution) results in distorted, inaccurate, and biased histories. The past is a complex place and generalizations that pass muster against the historical record are few. Take, for example, the common claim, often illustrated by supposedly historical stories about Darwin's life and work, that science and religion have always been in conflict. Such claims use the terms *science* and *religion* as though they are timeless categories, ignoring the fact that the definitions and boundaries between the two have changed in profound ways over time. Claims that science and religion are in harmony face similar problems since they too demand that science and religion be defined in timeless ways (generally how the proponent of harmony wishes them to be defined). But if we assume that the definitions, models, and boundaries that we believe in today existed in the past, we will completely misunderstand the hopes, fears, fights, and alliances that influenced those who lived in that past.[6]

Studying the Victorian era as historians also demands that we try to set aside (at least as an analytical guide) what we know about science based upon a subsequent century. The claim, for example, that belief in science increased because it had the power to save lives profoundly misrepresents the nineteenth century based upon twentieth-century developments. As we will see, ideas that most would probably call "scientific" (such as the idea that natural laws are uniform) gained power long before science could do anything to save lives or ameliorate suffering. Victorian stances on science must be explained *without* appealing to science's ability to do anything extraordinary at the bedsides of children. Remembering this fact is crucial to understanding the stances and beliefs of those who lived in the past, separate from present-day assumptions about the power of either science or religion.

Finally, doing good history demands that we pay close attention to the fact that, in a society highly stratified according to categories of class, religion, race, and gender, some got to write about suffering and some did not. Whoever had access to a printer's shop had a deep impact on the explanations that prevailed for both suffering and child mortality, especially the differences in child mortality that existed between populations. As we will see, the founders of what became Victorian science defined the development of "trustworthy" knowledge (and thus science) as the province of gentlemen. They defined gentlemen, in turn, as European, Christian, and upper-class. Most, but not all, of those who participated (in print form) in the Victorian debates over evolution were upper- or middle-class white men. Of course, these individuals had the power and privilege to write extensively about what they thought of everyone else (including what they thought everyone else thought). In doing so, they assumed it was right to use terms such as "savages" and "lower races" to indicate different placement on a "ladder of civilization" that they got to define. This assumption would have a profound influence on the fate of children throughout the British Empire. This is also a story, then, about whose suffering, and whose children, mattered most to those willing to declare what the world was like and why.

***

In teaching about Darwin for the past fifteen years, and thus in writing this book, I have depended upon the excellent scholarship of a number of historians and historians of science, but most especially the works of Peter Bowler, John Hedley Brooke, Janet Browne, Geoffrey Cantor, Matthew Day, Adrian Desmond, Paul Farber, John Greene, Piers Hale, Bernard Lightman, James Moore, Ronald Numbers, Diane Paul, Evelleen Richards, Michael Ruse, Suman Seth, Matthew Stanley, and James Turner. The approach and methodology of Paul Farber's course on the history of evolution theory, taught for years at Oregon State University, and Paul's fellow student Keith Bengtsson's at the University of Washington, pervade this book in countless ways. Thank you to Jonathan Kurtz, of Prometheus Press, for believing this version of the "Darwin Story" to be worth telling, to Arielle Lewis and David Bailey for their careful editing, and to the many family members, friends, and colleagues who have supported me in countless ways over the years. I am especially thankful to Erik Ellis, Thomas and William Ellis-Johnson, and Shauna and Chris Hansen, each of whom have, in different ways, connected this past to the present. This book is dedicated to my students, and it is written in loving memory of Claire and Henry Hansen.

Most of the primary sources consulted have been painstakingly preserved by an extraordinary resource for both learning and teaching about Darwin and his world: the Darwin Correspondence Project (cited as DCP in the endnotes, and available at www.darwinproject.ac.uk).

CHAPTER ONE

The Goodness of the Deity

IN THE WINTER OF 1697, THE ENGLISH NATURALIST AND ANGLICAN minister John Ray lost one of his four beloved daughters, Mary, to jaundice (acute liver failure). "My dear child," he wrote to his friend Sir Hans Sloane, "for whom I begged your advice, within a day after it was received, became delirious, and at the end of three days died apoplectic, which was to myself and wife a most sore blow." Thirteen-year-old Mary had been suffering from chlorosis, which turned her countenance green (now recognized as due to a deficiency of iron). Then her color turned from green to yellow and she became increasingly lethargic. Ray consulted a young physician and neighbor named Allen, who thought he had an "infallible cure." But Mary grew worse, and Ray and his wife had to try to comfort their child as she became delirious. To die "apoplectic" meant that in her final hours Mary was having violent convulsions. (Though neither Ray nor Sloane could know this, her jaundiced skin and eyes were produced by a buildup of bilirubin due to a malfunctioning liver, eventually leading to profound lethargy and delirium.)[1]

John Ray did not abandon his belief in God in the wake of this heartbreaking loss. Indeed, there is a hint in a letter written a month after Mary's death that he resolved to devote more, not less, time to his ministerial duties. For Ray and his British contemporaries, belief in the all-powerful, all-knowing, all-good God of Christianity was the foundation for understanding all of history, human nature, and the natural world. One individual's belief might look heretical to another. But the proposition that there is no God or that one cannot know one way or the other was virtually unthinkable. Atheists might haunt dreams, but in reality they were rare.[2]

How did one explain the suffering and death of a beloved child in this time and place, in which belief in the God who attended to the fall of every sparrow was so pervasive? And how did one obtain comfort amid such devastating loss? The aid provided by Christianity's emphasis on the prospect of heaven cannot be overestimated. Ministers emphasized heaven as a place of reunion with lost family members and friends, within "the loving care of Jesus." (Of course, as historian Hannah Newton notes, in the seventeenth century "the flipside to the belief in heaven was hell, a place that caused nightmares rather than pleasant dreams." That fact will eventually become crucial to our story.) With hope of heaven, the appropriate response to suffering, including the loss of a beloved child, was widely agreed upon, though it needed constant repeating: trust in God's will and a focus on what was required to ensure reunion in the hereafter.[3]

This agreement did not mean that resigning oneself to God's ways was easy. In 1660 Alice Norton recorded how, after the death of her son William, her daughter Naly admonished her for grieving so deeply. Surely, Naly demanded, her mother would not wish William back again, when "God has taken him to himself to heaven, where he has no sickness, but lives in happiness?" Norton expressed gratitude for the reproach, "and begged that the Lord would give me patience and satisfaction in his gracious goodness, which had put such words into the mouth of so young a child to reprove my immoderate sorrow for him." (Well aware another child might be taken, she also begged God to spare Naly "to me in mercy.")[4]

For much of the history of Christianity, Scripture provided the primary means of reinforcing the faith in God's goodness upon which Naly's words depended. But John Ray believed in an additional source of knowledge about God: the careful, meticulous study of nature. A few years before Mary's death, Ray wrote one of the most influential books of the so-called "Scientific Revolution" of the seventeenth century. Its full title was *The Wisdom of God Manifested in the Works of the Creation: The Heavenly Bodies, Elements, Meteors, Fossils, Vegetables, Animals (Beasts, Birds, Fishes, and Insects); more particularly in the Body of the Earth, its*

Figure, Motion, and Consistency; and in the admirable Structure of the Bodies of Man, and other Animals; as also in their Generation, &c. With Answers to some Objections. There Ray argued that the purposeful parts of animals and plants—from the human eye to the woodpecker's forked tongue—provided rational proofs of the existence of the wise, all-powerful, benevolent God of Christianity, the existence of a soul, and the promise of heaven. For John Ray (as for his contemporaries Isaac Newton and Robert Boyle, of Boyle's law fame), unbelief and doubt were not only heretical but *irrational.*

The tradition of natural history in which Ray worked would eventually have an enormous influence on the history of science, both in Britain and around the world. That tradition also tied a particular stance on the origin of species to what might be felt, thought, and done at the bedside of a sick child. To understand how and why this tradition became so powerful, we must immerse ourselves in the close ties drawn in the seventeenth century between Christianity and the study of nature, ties that governed British science for generations, including Darwin's.

The Wisdom of God

In calling upon *both* Scripture and nature in order to demonstrate the existence and character of God, John Ray was combining "natural theology" (the study of God based on the natural world) with "revealed theology," (the study of God based on Scripture). His belief in a mutually supportive relationship between natural and revealed theology depended upon an old assumption that God gave humans two books: the Book of Scripture and the Book of Nature. John Ray believed firmly that insights about God's existence and attributes could be obtained by the study of both "books." He peppered *The Wisdom of God* with evidence from Scripture to back up his conclusions regarding nature, then drew upon evidence from nature to back up his conclusions regarding the Bible.[5]

Traditionally, many Christian leaders had held that proofs about God's existence and character from nature were less important than demonstrations based on Scripture, revelation, and logic. Indeed, concern that, as a result of the "fall of man" (Adam and Eve's expulsion from the Garden of Eden as a result of their disobedience in eating from the Tree of Knowledge), human reason was illusory meant that natural theology remained subordinate to the Bible well into the seventeenth century. However, various political events in seventeenth-century Britain inspired some to turn to the Book of Nature as a crucial means of convincing individuals of God's extraordinary wisdom, power, and goodness. First, the Protestant Reformation (or Revolt, to Catholics) of the sixteenth century had divided Western European Christendom on some basic doctrines of faith. In Britain, King Henry VIII declared England Protestant, founding the Church of England (also known as the Anglican Church) in 1534. A century later, religious strife and contests for power between parliament and the monarchy resulted in the English Civil War. That conflict culminated in the beheading of the king in 1649 and an eleven-year "Commonwealth" (or "Interregnum") led by Oliver Cromwell. With no monarch and the Anglican "Church of State" abolished, a chaotic range of religious sects, each claiming to have discovered "true Christianity," spread throughout Britain.[6]

In 1660, the "Great Restoration" reestablished both a parliamentary monarchy and the power of the Anglican Church. That same year, the first scientific society in Britain was established: the Royal Society of London. Its members, including Isaac Newton, Robert Boyle, Robert Hooke (a founder of microscopy), and John Ray, had all lived through the chaos of the English Civil War. Each was committed to restoring social and political order. And each believed that to do so one must first establish consensus about God in a way that upheld the restoration of both the monarch and the Anglican Church. In asking, "How can one establish consensus about God, on grounds upon which all reasonable men might agree?" the members of the Royal Society of London argued that consensus could be achieved through the establishment of "matters of fact" via disciplined observation and experimentation. No longer could men (much less women) propose, as some of the Interregnum's religious sects had done, that intuition or feeling provides accurate knowledge about God. Clearly intuition and feeling differed among men and was thus untrustworthy as a basis for agreement. Men also differed regarding how to interpret Scripture. The founders of what they called the "New Sciences" argued that instead knowledge must be grounded in "facts" on which all rational men (and as we will see, they did mean just men) could agree.[7]

Now, because they believed (on biblical grounds) that God had created nature, the Royal Society men believed that consensus regarding the facts of nature would, in turn, lead to consensus regarding God. Indeed, they believed that correct knowledge about nature must be a crucial adjunct to the correct interpretation of Scripture. Isaac Newton, for example, saw his work on the universal law of gravitation as part of a divinely approved revelation of how God chose to govern creation. To make these arguments, the members of the Royal Society embraced a long-standing tradition of distinguishing between the first cause (God) and secondary causes (the natural laws instituted by God, as legislator of the universe) of natural phenomena. Although Royal Society members did much of their work within the realm of secondary causes, they insisted that in doing so they were studying the means through which God, the first cause, generally chose to govern. They also made two fundamental assumptions about what constitutes an appropriate secondary cause: First, to varying degrees Newton, Boyle, Ray, and company thought of nature as analogous to a machine (rather than as analogous to an organism). In this view the key to understanding nature is to analyze its parts and study how those parts work together to perform certain functions. Second, they adopted atomic theories of matter—that is, they assumed matter is made up of indivisible units (rather than, say, four elements) whose behavior determines all we see, hear, touch, and feel. (Each of these assumptions still governs science above the quantum level.) In the parlance of the day, this new "mechanical philosophy" was united with the Copernican view of the cosmos and designated "the new sciences." (A century later, proponents used the term *Scientific Revolution* to describe these transformations.)

It turns out that a profound danger lurked within these moves, a danger of which the seventeenth-century proponents of the new sciences were well aware. Atomic theories of matter were associated with the "atheistical" views of ancient Greek philosophers, who held all of creation to be the result of chance, or (in the language of the day) a "fortuitous concourse of atoms." Might those who espoused atomistic theories of matter in the seventeenth century also deny God's role in creation?

5

Given the Royal Society members' strong interest in helping to restore social order (*all* assumed atheism would result in social chaos), this was a serious problem indeed. Historian John Hedley Brooke describes the dilemma facing the Royal Society founders as follows:

> *What role could be left for God to play in a universe that ran like clockwork? Would one have to side with those who became known as "deists," who restricted that role to the initial creation of a law-bound system, and who attacked Christian conceptions of a subsequent revelation? Would God's special providence, His watchful concern for the lives of individuals, not be jeopardized if all events were ultimately reducible to mechanical laws?*[8]

In other words, were these proponents of the new sciences sealing the argument for theism (the close personal God of the Old and New Testaments), or were they at best proving the existence of a more distant, *deist* God, who created the universe and retired, letting natural laws complete all additional creative work?

The Royal Society members knew that, across the channel in Holland, Baruch Spinoza, having rebelled from the Jewish beliefs of his family, was drawing upon the new sciences to criticize the idea of a providential, personal God. Spinoza adopted a "necessarian" (all effects necessarily follow from causes) view of the universe in which God governed solely via natural laws. In Britain, a nation in which calls for obedience to God's will in the face of affliction were absolutely dependent on belief in a close personal God who attended the fall of every sparrow (from the joys and miseries of individual human lives to the power of the king and Anglican Church), securing God's personal, providential rule in the face of such radical ideas was essential.[9]

Here is the argument the founders of the Royal Society devised to prevent the study of natural law from distancing God from creation: machines (which by definition have purpose) can neither make nor move themselves. They require a designer (the mechanic, machine-maker, or engineer) who, having imagined a purpose (such as telling time), has contrived how to form a machine (say, a clock). Having adopted the mechanical philosophy, the proponents of the new sciences imagined natural phenomena as analogous to machines. Any purpose in nature thus became an argument for the existence of a Designer.

John Ray (who we met at the beginning of the chapter) applied this argument to natural history, carefully documenting the extraordinarily purposeful parts of animals and plants: the woodpecker's feet, the snake's scales, the owl's eyes. His most famous and influential book, *The Wisdom of God Manifested in the Works of the Creation*, was central to the Royal Society's counterattack against anyone using the mechanical philosophy and new sciences to defend deism or atheism (or, worse, accuse Royal Society members of doing so). Throughout its pages, Ray reveled in detailed descriptions of animal and human anatomy that showed "the exact fitness of the Parts of the Bodies of Animals to every one's Nature and Manner of Living." He carefully described how woodpeckers "have a Tongue which they can shoot forth to a very great length, ending in a sharp stiff bony Rib, dented on each side; and at Pleasure thrust it into the Holes, Clefts, and Crannies of Trees" to draw out insects. "More over," he continued, "they have short, but very strong Legs and their Toes stand two forwards, two backwards . . . very convenient for the climbing of Trees, to which also conduces the stiffness of the Feathers of their Tails, and their bending downward, whereby they are fitted to serve as a Prop for them to lean upon, and bear up the Bodies." The only rational explanation of such perfect, purposeful design, he argued, was that species were created by a "Designing Mind." Given this fact, Ray found it quite "a wonder" that "there should be any man found so stupid and forsaken of reason as to persuade himself, that this most beautiful and adorned World was or could be produced by the fortuitous concourse of Atomes" (that is, by chance). When later naturalists (from German anatomists to British geologists) updated Ray's work throughout the eighteenth and nineteenth centuries, they added new natural history facts and details, but the argument remained essentially the same: the purposeful parts of animals and plants demonstrate the existence of an all-wise, all-powerful, all-benevolent (omniscient, omnipotent, and omnibenevolent) Designer.[10]

Ray's belief that the usefulness of parts and behaviors would convince his readers of God's close, attentive governance over creation meant he searched for the purpose and meaning of every natural fact. When he came to human anatomy, Ray described the purposeful parts of the body as beautiful examples of the effect of wisdom and design. Thus, he concluded, the body of man was "proved to be the Effect of Wisdom because there is nothing in it deficient, nothing, superfluous, nothing but hath its End and Use." God even, in his great benignity, bestowed two hands, eyes, nostrils, ears, feet, and breasts, so that if one should be disabled all would not be lost. Sometimes Ray had to concede defeat in the search for the purpose of a part, at least temporarily. "Only it may be doubted," he wrote, "to what use the Paps [nipples] in Men should serve." Perhaps for ornament, perhaps for conformity between the sexes, "and partly to defend and cherish the Heart." "However," he insisted, "it follows not that they or any other Parts of the Body are useless because we are ignorant." Ray did concede that a body made by chance "must in all Likelihood have had many of these superfluous and unnecessary Parts," the demonstration of which might give "some Pretence to doubt whether an intelligent and bountiful Creator had been our Architect." But he was confident that the purpose of organs that seemed superfluous would one day be found through careful research. Meanwhile, this tradition focused naturalists' attention on the fitness of form to function and inspired generations of naturalists (entomologists, ornithologists, botanists, etc.) to amass encyclopedias of purposeful parts, all confidentially described as testaments to God's wisdom, power, and goodness.[11]

FOR THE RELIEF OF MAN'S ESTATE

Although a surprise to those used to tales of constant warfare between science and religion, this tradition of bolstering Christianity with science became an influential route by which the following assumptions spread: First, that we can, through disciplined methods, figure out what nature is really like; second, that natural objects can be understood by comparing them to machines, composed of "matter in motion"; third, that the use of observations and experiments to establish "matters of fact" is the best means of establishing true knowledge about nature; and fourth, that improved knowledge is an important means of ameliorating future suffering.

We should spend some more time with the fourth assumption, for it will be particularly important to our story. Few Christians held that the belief that God governed all meant that one should do nothing in the face of suffering. Christ's statement that "the poor will always be with you" was not an injunction to withhold charity, but accompanied the command to "open your hand wide to your brother." (Both declarations were originally from the Old Testament.) Strong precedents existed for combining a pious belief in God's providence with a commitment to action to ameliorate suffering. After all, Protestant, Catholic, Jewish, and Islamic religious traditions each had robust medical traditions going back centuries. In the fourth century St. Basil the Great argued that, although the fall of man had resulted in both death and disease for all of Adam and Eve's descendants, a merciful and benevolent God had provided "medical art" to "relieve the sick, in some degree at least." The Christian must simply trust that God had a good reason when medicine failed, and remember that ultimately the fate of the soul, not the body, was what mattered most. As one medieval Christian wrote, the good physician must remember that "either [medicines] would cure or the patient 'will die for God did not will that they should be healed.'"[12]

Proponents of the new sciences such as John Ray expanded on this tradition by arguing that through their "new" disciplined methods, they could develop the knowledge to ameliorate *more* suffering in future. To do so they drew upon the biblical interpretations of an English statesman named Francis Bacon. Like most Christians, Bacon believed that suffering had entered God's creation at the fall of man. But Bacon also held that the story of the fall explained man's *ignorance* of nature, including its medicines. After all, Genesis proclaimed that Adam had once known the names of all animals and plants. While Bacon agreed with the prevailing view that the fall had disconnected human perception from reality, he also believed that God had provided a very specific path through which certain men might recover knowledge and, in doing so, prepare the earth for Christ's promised Second Coming and heavenly reign. That path was the careful observation and experimental study of nature. Through disciplined methods, Bacon argued, man might divest his interpretation of nature of "Four Idols of the Mind": our tendency to rely too much on sense perception, overgeneralize, and ignore counterevidence; our tendency to agree with ideas we prefer given our particular perspective; our tendency to be misled by the limitations of language; and our propensity to be caught up in our own preconceived ideas about the world.

Francis Bacon imagined humans *participating* in the prophesied movement toward the Kingdom of God by actively recovering knowledge lost during the fall of man. To secure true knowledge, he argued, men must rely on their senses, aided by new instruments like the microscope and telescope and disciplined by observation and experiment. Better knowledge would in turn allow humanity to reproduce the Garden of Eden, a place characterized by an absence of suffering, disease, and even death. (For Bacon, Eve had sinned not in being curious, but in seeking wisdom through the route of eating from the Tree of Knowledge, rather than through the laborious path of finding out for herself.)[13]

Forty years later, the Royal Society modeled itself on Bacon's proposals. It embraced Bacon's promises that the years prior to Christ's return would be paved with radical new innovations and the *amelioration* of suffering on Earth. Assured that their endeavor was pious and Christian, British proponents of the new sciences envisioned humanity taking up the tools provided by a wise and good God to learn new things in order to ameliorate suffering. They also held that the primary means of producing knowledge must be the cooperative, organized efforts of "learned Christian gentlemen."[14]

Drawing upon Bacon's interpretations of the relation between the fall of man and human ignorance allowed the Royal Society members to see that ignorance as *temporary*, to be combatted via the study of the natural laws instituted by God. John Ray, for example, wrote that God had provided seeds and fruits useful for food and medicine, capable "of being meliorated and improved" by human art. God had also made man a sociable creature "for the improvement of thy Understanding by Conference, and Communication of Observations and Experiments." In a revised and expanded edition of *The Wisdom of God*, Ray described plants as demonstrating "the illustrious Bounty and Providence of the Almighty and Omniscient Creator, towards his undeserving Creatures," and listed plants like the Jesuit's bark tree, the poppy, the rhubarb, the jalap, and others with "uses in Curing Diseases." (The Jesuit's bark tree gives quinine, the poppy gives opium, and rhubarb and jalap are both strong purgatives. All were obtained from overseas.) This was not necessarily a long list, but Ray was sure "there may be as many more as yet discovered, and which may be reserved on purpose to exercise the Faculties bestowed on Man, to find out what is necessary, convenient, pleasant or profitable to him." After all, God had also made man an adventurous creature with a desire to see unknown lands and bring home "what may be useful and beneficial." (Notice the implicit claim of divine approval of European exploration and colonial expansion.)[15]

Like Bacon, John Ray assumed that a benevolent God had passed a certain amount of agency to humans for the purpose of alleviating suffering. He envisioned pious men taking up the tools provided by God (including reason) to improve the human condition, even as he firmly believed one must bow before the inscrutable wisdom of God's ways when medicine and human knowledge failed. We can thus make sense of the fact that, amid his prayer-filled letters, in which appeals to God to deliver himself and his friends from illness were constant, there is also palpable hope that he or his friends might, with God's approval, find the proper medicine to take away the pain.

There is evidence, even in the seventeenth century, that this could be a hard bargain, full of regret for action not taken because one did not (yet) know enough. John Ray blamed himself, for example, when he failed to give the right medicine to his daughter. "Nothing afflicts me so much," he wrote shortly after, "as that I did [not] in time make use of that remedy, which had proved so effectual to my own relief and cure in the same disease." This tension, as individuals tried to balance belief in God's providence with the possibilities of human agency, will be central to our story.[16]

The Royal Society members' claims that they could both improve knowledge and ameliorate suffering did have their critics. All were outsiders to its "club" meetings and publications. One of those critics, Margaret Cavendish, became the most published woman in the seventeenth century, yet she was never allowed to become a member of the society. (The rule barring women was not removed until 1945.) As historians of science have demonstrated, European traditions of science grew directly out of a monastic and clerical culture that excluded women. A consensus that God commanded the husband to rule over the wife, designed women to bear children, and made women "by nature" emotional rather than rational meant that proponents of the new sciences did not think women could even *do* science. Furthermore, the particular system the Royal Society developed to produce "matters of fact" (trusting each other, for example, to produce honest accounts of experiments since not all experiments could be replicated) meant that those stereotyped as beguiling and untrustworthy could not participate in the production of natural knowledge. Royal Society members saw "trustworthy" knowledge (including statements about both nature and God) as the province solely of "gentle*men*." An exceptional wealthy woman could serve as both patron and inspiration, but she could not show up at a Royal Society meeting and be treated as an intellectual equal.

On the other hand, as an outsider, Margaret Cavendish did not need to follow the Royal Society's rules of either manners or methods. Cavendish deftly criticized the Royal Society's claims that they could produce both true and useful knowledge. (Note proponents of the new sciences had obtained royal patronage by arguing their work was both useful and pious.) She called the Royal Society members "Boys that play with watery Bubbles . . . worthy of reproof rather than praise, for wasting their time with useless sports,"[17] and demanded to know whether their little microscopes that revealed the wondrous details of lice anatomy could hinder the tiny creatures from biting! It was, at the time, a fair point. The Royal Society did not deliver on its promises to substantially improve the human condition for nearly two hundred years. (Scientific knowledge about lice wouldn't save lives until the 1910s.)[18]

Cavendish's critique is a good reminder that, although it is tempting to assume that what we now know as physics, chemistry, and biology became powerful because the Royal Society's methods did, as promised, ameliorate suffering via new knowledge, such an assumption is not justified by the historical record. In fact, for generations the new sciences gained power due to the utility of the arguments they offered about God's providence, rather than their ability to either improve medicine or save lives. Child mortality rates, one of today's most important indexes of a society's ability to ameliorate suffering, did not initially begin dropping because of anything the Royal Society was up to, much less because of new medical techniques. The one discovery that did prevent child mortality, inoculation against smallpox, first appeared in Royal Society publications as accounts of African and Ottoman medical traditions, rather than new insights into nature provided by the new sciences. This fact did not prevent proponents of the Royal Society's methods, like Thomas Jefferson, from using smallpox inoculation as evidence of the superiority of European science in their interactions with other (especially Indigenous American) ways of thinking about the world. But for generations, it is *hope* of a (divinely approved) "relief of man's estate," rather than clear evidence of success, that must be tracked as the new sciences spread.

One of the most profound illustrations of both the close relationship between Christianity and the study of nature and individuals' complex negotiations between belief in God's close personal governance and action to ameliorate suffering may be found in the work of Phillis Wheatley (we do not know the name her parents gave her). Kidnapped from her home in West Africa at the age of seven and taken to the British colonies in New England to be sold into slavery, Wheatley was purchased by the Wheatley family. In an uncommon (and eventually illegal) move, the Wheatleys taught her to read and write. When the family discovered she was a poet they took her to England, where the evangelical abolitionist community embraced her as evidence of Africans' common humanity. (Given, as we will see, the use of both Scripture and the new sciences to argue otherwise, the point had to be explicitly made). In her poems, Wheatley called upon both faith and reason to demonstrate God's all-powerful, benevolent, and wise providence in the wake of great affliction. Her "funeral poems," composed for families whose children had died, urged that devout parents take refuge in the promise of heaven, where there would be eternal reunions and an end to all affliction. Wheatley imagined bereaved parents falling "prostrate, withered, languid and forlorn," asking whence their child flies, and then admonished:

Cease your complaints, suspend each rising sigh,
Cease to accuse the Ruler of the sky. . . .
He brought that treasure which you call your own.
The gift of heaven entrusted to your hand.
Cheerful resign at the divine command:
Not at your bar must Sovereign Wisdom stand.[19]

Though most of her appeals rely on scriptural promises to remind her readers of heaven and God's providence, Wheatley also turned to God's other book, the Book of Nature, as evidence of His goodness. She wrote of the "wondrous works" of the Almighty as reflected in the cosmos, trees, flowers, and human frame, from which "What Power, what Wisdom, and what Goodness shine!"

The Atheist sure no more can boast aloud,
Of chance, or nature, and exclude the God;
As if the clay, without the potter's aid,
Should rise in various forms and shapes self-made,
Or worlds above, with orb o'er orb profound,
Self-moved, could run the everlasting round.
It cannot be—unerring Wisdom guides
With eye propitious, and o'er all presides.[20]

Wheatley had faith, based on both nature and Scripture, that God ruled over all. She believed in Scripture's message that all suffering is redemptive and that God brought good from all evil, even slavery: "'Twas mercy brought me from my *Pagan* land," she wrote, for thus she had learned of God and of Christ.[21] Recall that Wheatley was immersed in a worldview in which God *must* have a reason for permitting such tremendous affliction, and in which to imagine otherwise—that so much suffering had no purpose, or that it arose from an evil power that was more powerful than God—was virtually unimaginable, and if imagined, heretical. But for Wheatley (as for abolitionists and physicians) belief in God's will did not mean one must not fight against suffering. Finding God's hand in her own history did not mean Wheatley believed one must submit to what sin-ridden humanity did on Earth. Although she believed that Christians must resign themselves to God's will when children died, Wheatley made very clear that slavery cruelly demanded parents resign themselves to something *humans* did. It was a human soul "by no misery moved," she wrote, "That from a father seized his babe beloved."[22] Wheatley did not believe that faith in a close personal God meant that the state of the earth could not be questioned. Within Wheatley's version of Christianity, humans, as fallen creatures, were sinners, yet they must try to choose good over evil, for *all* God's children.[23]

THERE CANNOT BE DESIGN WITHOUT A DESIGNER

As far as we know, Charles Darwin read neither John Ray's *The Wisdom of God* nor Phillis Wheatley's poems. But he did read—and by his own account knew by heart—an influential update of their argument that nature demonstrated the truths of Scripture. Composed by an Anglican clergyman named William Paley, the book appeared in 1802 and was entitled *Natural Theology: or, Evidences of the Existence and Attributes of the Deity, Collected from the Appearances of Nature.* Darwin later described how he was "charmed and convinced by the long line of argumentation" in Paley's works.[24]

Paley's book began with what would become one of the most famous passages in nineteenth-century natural history. "In crossing a heath," he wrote, "suppose I pitched my foot against a stone, and were asked how the stone came to be there." One would no doubt conclude that the stone had lain there for ages. But if, by contrast, he had found a watch upon the ground one would, of course, give a different answer, namely that the watch had been made by a person. But why the different answers? Paley replied:

> *For this reason, and for no other, viz. that, when we come to inspect the watch, we perceive (what we could not discover in the stone) that its several parts are framed and put together for a purpose, e.g., that they are so formed and adjusted as to produce motion, and that motion so regulated as to point out the hour of the day. . . . This mechanism being observed . . . the inference, we think, is inevitable, that the*

watch must have had a maker: that there must have existed, at some
time, and at some place or other, an artificer or artificers who formed
it for the purpose which we find it actually to answer; who compre-
hended its construction, and designed its use.[25]

This imaginary reflection on what a person must conclude about
the origin of a lost watch prefaced more than five hundred pages of
detailed descriptions of human, animal, and plant anatomy and behavior.
Throughout, Paley compared the purposeful parts of living beings (from
human hands to woodpecker's tongues) to various human-built machines,
to argue that organisms, like machines, must have been designed by an
intelligent being (a grand Watchmaker).

Paley's first and most famous example (and the one that would eventually give Charles Darwin a great deal of trouble) was the human eye:
"As far as the examination of the instrument goes, there is precisely the
same proof that the eye was made for vision, as there is that the telescope
was made for assisting it. They are made upon the same principles; both
being adjusted to the laws by which the transmission and refraction of
rays of light are regulated." Every part testified "counsel, choice, consid-
eration, purpose," since none could be accounted for by any other means
than by design: "There cannot be design without a designer," Paley
argued, "contrivance without a contriver; order without choice; arrangement, without anything capable of arranging; subservience and relation
to a purpose, without that which could intend a purpose."[26]

Paley updated the Royal Society's design argument for a nation
reaping the benefits, and sowing the tragedies, of the Industrial Revolution. While John Ray's book was a product of late-seventeenth-century
sensibilities and concerns, and was peppered with biblical passages, Paley
quoted Scripture relatively rarely. He also compared living organisms
to new machines beyond anything John Ray could have imagined. He
appealed to his readers' increasing familiarity with "watches, telescopes,
stocking-mills, steam-engines," to demonstrate that, while reproduction, generation, and development proceeded via secondary causes that
could be studied via the methods of science, all human experience with

purposeful parts showed that both species and adaptations occurred via independent, creative acts of the first cause, God.[27]

As we will see in the next chapter, another explanation of the origin of species had been proposed by the time Paley wrote *Natural Theology*. In France and in England, radical books claiming that species arose via something called "transmutation" (an earlier word for evolution) had appeared. Paley had scriptural reasons to be suspicious of such claims. After all, the conclusion that God created species supported a relatively straightforward interpretation of Genesis (in which interbreeding populations, or species, are called *kinds*): "And God said, 'Let the earth bring forth the living creature after his kind, cattle, and creeping thing, and beast of the earth after his kind': and it was so" (Genesis 1:24). Based on a literal interpretation of "bring forth" and "kind," God created species independently; they did not arise through some materialistic process from other forms.

Paley believed (as had Isaac Newton, Robert Boyle, and John Ray) that biblical passages might need to be reinterpreted based on new discoveries about nature. This was a standard interpretative rule going back to St. Augustine in the fifth century: as divine creations, nature and the Bible cannot contradict one another. If a *demonstrated* truth about nature seemed to contradict something in the Bible, then the relevant passages must be reinterpreted. Indeed, nearly two centuries earlier, Galileo had cited this rule in his defense of a sun-centered cosmos. He got into trouble because, while he had good arguments in favor of a sun-centered cosmos, he did not have demonstrative proof, and, to make matters worse, he offered his own biblical interpretation at a time in which the Catholic Church had issued strict rules against laymen interpreting Scripture on their own. Meanwhile, for Protestant Christians, a lack of centralized control and the fundamental premise that "two truths cannot contradict" had encouraged a more allegorical reading of the scriptural passage in which Joshua commanded the sun, rather than the earth, to stand still in the sky.[28]

No naturalists believed that a *scientific* rationale existed for interpreting the line "bring forth the living creature after his kind" in Genesis to mean anything other than the creation of fully formed species. Indeed, when Paley did address the idea that one species could change into another, he was able to emphasize purely rational, rather than scriptural, arguments. Speaking of the extraordinary design of the elephant's trunk, which so beautifully compensated for the animal's short, unbending neck, Paley quickly and easily dispensed with the idea that species could somehow change themselves into other forms:

> If it be suggested, that this proboscis may have been produced in a long course of generations, by the constant endeavour of the elephant to thrust out his nose . . . I would ask, how was the animal to subsist in the mean time, during the process; until this prolongation of snout were completed? What was to become of the individual, whilst the species was perfecting?[29]

Paley did not write much about such theories, however. There was no reason (in 1802) to take evolutionary views seriously as scientific propositions. "There is no foundation whatever for this conjecture," Paley insisted, "in anything which we observe in the works of nature: no such experiments are going on at present; no such energy operates as that which is here supposed, and which should be constantly pushing into existence new varieties of beings." For a species to change into another species seemed to require that matter act of its own accord. Yet how could a clock rearrange its material particles into a different kind of clock? (Variation in domesticated animals such as dogs was not seen as relevant to the argument since first, such breeds do not exist in a state of nature, and second, although breeders had altered domestic animals by selecting for certain traits, no breeder had created a new species or organ.)[30]

Paley cited the most famous anatomists of the day in support of his argument. Drawing upon the work of William Cheselden, for example, he compared human and animal skeletons, muscles, and circulatory systems (including the "engine which works this machinery," the heart and blood vessels) to grand machines, with God as the Great Engineer. "For my part," he wrote, "I take my stand in human anatomy." Furthermore, Paley held that these examples of mechanism—in the pivot on which the head turns, the pulley system in the muscle of the eye, the knitting of the intestines to the abdominal wall—provided obvious and convincing evidence of the Designer's close attention to creation. Here is Paley's description of the epiglottis, which keeps food from blocking the windpipe:

> Both the weight of the food, and the action of the muscles concerned in swallowing, contribute to keep the lid close down upon the aperture, whilst any thing is passing; whereas, by means of its natural cartilaginous spring, it raises itself a little, as soon as the food is passed, thereby allowing a free inlet and outlet for the respiration of air by the lungs. Such is its structure: And we may here remark the almost complete success of the expedient, viz. how seldom it fails of its purpose, compared with the number of instances in which it fulfills it. Reflect how frequently we swallow, how constantly we breathe. In a city-feast, for example, what deglutition, what inhalation! yet does this little cartilage, the epiglottis, so effectually interpose its office, so securely guard the entrance of the wind-pipe, that whilst morsel after morsel, draught after draught, are coursing one another over it, an accident of a crumb or a drop slipping into this passage (which nevertheless must be opened for the breath every second of time), excites in the whole company, not only alarm by its danger, but surprise by its novelty. Not two guests are choked in a century.[31]

Paley also dealt with parts for which no purpose could yet be discerned. John Ray had admitted that the presence of "superfluous and unnecessary Parts" might give "some Pretence to doubt whether an intelligent and bountiful Creator had been our Architect." But like Paley, he was absolutely confident that the purpose of organs that seemed unnecessary would one day be found through careful research. They were both stumped, however, by male nipples. "I confess myself totally at a loss to guess at the reason," Paley confessed, "for this part of the animal frame." But in contrast to Ray, Paley argued that even purposeless parts could not vitiate the conclusions drawn from parts for which the purpose is known.[32]

Paley's book also included a detailed survey of comparative anatomy (the study of the similarities and differences between different kinds of animals). After describing the structure of human eyes in great detail, he then compared them to the eyes of fishes, eels, birds, and mammals in order to demonstrate God's wisdom in adjusting a basic plan to different needs. "Thus," Paley concluded, "in comparing together the eyes of different kinds of animals, we see, in their resemblances and distinction, one general plan laid down, and that plan varied with the varying exigencies to which it is to be applied." He also included descriptions of what he called "peculiar organizations" (such as the marsupial's pouch, the camel's stomach, and the woodpecker's forked tongue) and "prospective contrivances" (such as milk ducts). All, he argued, demonstrated the existence of the all-wise, all-powerful, all-benevolent God who attends to the fall of every sparrow:

> *Under this stupendous Being we live. Our happiness, our existence, is in his hands. All we expect must come from him. Nor ought we to feel our situation insecure. In every nature, and in every portion of nature, which we can descry, we find attention bestowed upon the minutest parts. The hinges in the wings of an earwig, and the joints of its antennae, are as highly wrought, as if the Creator had nothing else to finish. We see no signs of diminution of care by multiplicity of objects, or of distraction of thought by variety. We have no reason to fear, therefore, our being forgotten, or overlooked, or neglected.*[33]

Here Paley wrote about the hinges of the earwig rather than falling sparrows, but the point was the same: looking at nature rightly, rationally, and piously, one would be reminded that nothing happened without God, all affliction had a divine purpose and meaning, and humans must trust that all would come right in the end.

WHEN THE WATCH GOES WRONG

No doubt it was easy to see wisdom, power, and goodness in the universal law of gravitation, a perfectly functioning pair of eyes, a productive landscape, a healthy child, the birth of an infant, and many other seemingly benevolent provisions. But what of kidney stones, jaundice, blindness, and a whole host of diseases that cut off infants in their first few hours or days? What of smallpox and so-called "monstrous" births? What of aldermen who, on occasion, choked on their beefsteak? And did not some aspects of nature imply that the Creator is unconcerned with creation, or had even on occasion bungled His inventions?

Like all earnest ministers, Paley faced a dilemma: How to support individuals' faith in God's goodness amid constant suffering? Paley clearly felt the suffering of his parishioners keenly. During an economic depression he wrote of "the disappointments and distresses, the changes and failures, which the disturbed state of commerce hath lately brought upon those who are engaged in it." He had seen individuals suffering "from dread and anxiety, if not from actual losses and privations."[34]

The titles of some of the prayers Paley included in his work, *The Manner of Visiting the Sick; or, The Assistance that is to be given to sick and dying persons by the ministry of the clergy*, addressed dozens of additional afflictions to which a minister might be called:

A Prayer for one who is dangerously ill
A Prayer for a Woman who cannot be delivered without
Difficulty and Hazard
A Prayer for Grace and Assistance for a Woman after Delivery,
but still in Danger
A general Prayer for Preparation and Readiness to die
A Prayer for one who is troubled with acute Pains of the Gout,
Stone, Colic, or any other bodily distemper
A Prayer for a Person in a Consumption, or any lingering
Disease
A Prayer for a Person in the Small-Pox, or any . . . ravaging
infectious Disease
A Prayer for One that is Bed-ridden
A Prayer for a Lunatic
A Prayer for One under deep Melancholy and Dejection of
Spirit

A Prayer for One under the dread of God's Wrath and everlasting Damnation[35]

Taken together, the list captured ubiquitous human afflictions that had inspired some to question whether God—or at least the Anglican God to whom these prayers were addressed—even existed.

Paley knew from personal experience how, as he wrote toward the end of *Natural Theology*, occasions would inevitably arise in life that would try the firmness of "our most habitual opinions" about God.[36] He was forty-eight when his wife of fifteen years, Jane Hewitt, died in 1791. He had left a teaching position at Cambridge to marry her (professors were the only clergymen for whom vows of celibacy were required) and had become a widower with eight children. We do not have detailed knowledge of Paley's family life, but by all accounts, he adored his children. He once wrote that he saw the benevolence of the Deity "more clearly in the pleasures of very young children than in anything in the world. . . . Every child I see at its sport affords to my mind a kind of sensible evidence of the finger of God and the disposition which directs it." He and Jane, like so many parents, lost at least (not all births were registered) two infants: John and Francis.[37] Were occasions like this on Paley's mind when he wrote that when faith is tested, "it is a matter of incalculable use to feel our foundation; to find a support in argument for what we had taken up upon authority"? Clearly, Paley hoped that *Natural Theology*, which appealed to readers' observation and reason, might provide that crucial additional foundation. (He composed his concluding chapters—including the claim that "it is a happy world after all"—while suffering from intestinal cancer, and writing his odes to the perfect design of human anatomy as his own machine was painfully breaking down.)[38]

Natural Theology was thus not some naive portrait of the world constructed by a man who had never experienced or witnessed suffering. Daily experience with disease, suffering, and death runs through the text. He knew, for example, the purpose of many parts by virtue of what happened when those parts broke (cartilage by virtue of the grinding pain produced when it was gone; tooth enamel by the fact one could not eat without its protective cover). Clearly, one had to admit the existence of pain and suffering. But that by no means vitiated the argument: indeed, the whole point of *Natural Theology* was that the study of purposeful parts proved God's close benevolence, wisdom, and goodness, *despite* such suffering.

In his sermons, Paley emphasized that afflictions like illness must be seen as arising directly from the first cause, God, for one's benefit. "The minister should frequently be exhorting the sick man to patience," Paley advised, "and a blessed resignation to the will of God." No individual, he wrote, should look upon his sickness as the effect of secondary causes, but rather "as inflicted on him by Divine Providence for several wise and good ends." Those ends might include a trial of one's faith, the exercise of patience, the punishment of sins, the amendment of one's life, or an example to others (assuming the patient practiced both resignation and reform as a result of the illness). Or it might be "for the increase of his future welfare, in order to raise him the higher in glory hereafter, by how much the lower he hath been depressed here." For women "with child" or "in travail" (pregnant or in labor), the minister must pray that the sufferer was not disquieted by the fear of any evil, since none could happen to her without God's permission.[39]

Such prayers depended, of course, on faith that God's governance is trustworthy and good. Though as a good Anglican Paley included warnings that human beings deserved eternal torment as a reminder of God's justice, he emphasized the promise of salvation as a reminder of God's mercy and benevolence. The minister faced with an individual depressed by fear of eternal damnation, for example, must remind the afflicted not to distrust God's goodness, and that the mercies of God are infinite. Thus, in sermons and prayers, Paley insisted that affliction, from the travails of childbirth to the loss of beloved children, must not be viewed as the result of secondary causes. All must be viewed as inflicted by "Divine Providence" for wise and good ends.

Deep within the pages of *Natural Theology*, however, an old dilemma lurked. Paley was discussing "the origin of evil" when he proposed that one of the most comprehensive solutions to the question of why so much suffering existed entailed a focus on "general rules" (by which he meant "secondary causes" or "natural laws"):

We may, I think, without much difficulty, be brought to admit the four following points: first, that important advantages may accrue to the universe from the order of nature proceeding according to general laws: secondly, that general laws, however well set and constituted, often thwart and cross one another: thirdly, that from these thwartings and crossings, frequent particular inconveniencies will arise: and, fourthly, that it agrees with our observation to suppose, that some degree of these inconveniencies takes place in the works of nature.[40]

In other words, Paley noted that much suffering inevitably arose from the need for natural laws (which were great and divinely designed goods) to be uniform and consistent. But this "solution" raised a profound question: What had happened to God's close personal attention that ensured a divine origin, purpose, and meaning for every affliction?

As we have seen, the seventeenth-century founders of the new sciences knew that emphasizing natural laws might distance God's hand too much. They accepted that risk for various reasons. First, they firmly believed that, rightly and piously studied, God's governance via uniform natural laws provided a means of divinely approved amelioration of human suffering. For only then could an observant physician learn the regular course of disease and discern when a helping hand was needed. Second, they were certain that the study of the purposeful parts of animals and plants would close (rationally, logically, and scientifically) the distance to which God might be driven by the study of natural laws. William Paley embraced both these commitments, and he brought them into the nineteenth-century study of both natural history and anatomy.

Critics, in the seventeenth century and Paley's day, demanded to know in what sense God was present, if both disease and healing could be described as the result of "a certain assemblage of natural causes." The problem was this: If one could describe the universe as governed by a delegated power (or secondary cause), why did one need the Delegator (the first cause) in the first place? And what then became of God's close attention to each sparrow's fall? Or a beloved child's illness? Indeed, some accused Paley of deism (belief in a universe guided by natural laws set in motion by a distant God). Such critics either forgot or distrusted *Natural*

Theology's long passages about God's close attention to his creation. For it is important to note that Paley describes "general laws" as sometimes leading to "apparent evil, for which *we* can suggest no particular reason." He was absolutely confident that the resulting "inconveniences" (i.e., incidents that caused suffering) were still purposeful and directed by God, who planned, ordained, and knew all and must be trusted for having designed the world just so.[41]

Paley did try to answer those who, as we will see, pointed out that "perfectly designed" infant-machines sometimes tragically malfunctioned. In considering a watch, he argued, the demonstration of design would not be invalidated if "the watch sometimes went wrong" or "seldom went exactly right." The purpose of the machinery, the design, and the designer would still be evident "in whatever way we accounted for the irregularity of the movement, or whether we could account for it or not." Indeed, he argued that disease could even be placed on the "good" rather than the "evil" side of the ledger when one considered that, first, human sympathy (a great good) grew amid bodily suffering; second, disease reconciled mankind to death; and third, suffering and pain inspired greater appreciation for health when it was present. "A man resting from a fit of the stone or gout," Paley wrote, "is, for the time, in possession of feelings which undisturbed health cannot impart. They may be dearly bought, but still they are to be set against the price." (As we will see, some argued that child suffering and death did not at all seem accounted for by such arguments.)[42]

Ultimately, Paley argued that, on balance, happiness outweighed misery and pleasurable sensations were more common than miserable ones. Surely these facts proved, he was sure, the existence of a benevolent Creator who intended human happiness, if not on Earth, then surely in heaven. Confident he had made his case secure, he concluded:

> *It is a happy world after all. The air, the earth, the water, teem with delighted existence. In a spring noon, or a summer evening, on which-ever side I turn my eyes, myriads of happy beings crowd upon my view. "The insect youth are on the wing." Swarms of newborn flies are trying their pinions in the air. Their sportive motions, their wanton*

mazes, their gratuitous activity, their continual change of place with-
out use or purpose, testify their joy, and the exultation which they feel
in their lately discovered faculties.[43]

Natural history, in Paley's world, was supposed to strengthen individuals' faith that their afflictions (e.g., sickness, misery, the death of a child)
were also somehow part of God's loving care, even as naturalists, natural
philosophers, anatomists, and physicians tried to understand the secondary causes through which God governed. One might explain a sparrow's
fall via the universal law of gravitation, but the wonderful, purposeful
design of the sparrow's feet, beak, hollow bones, feathers, and wings
proved the Great Legislator, God, was still present. Meanwhile, the disciplined study of natural laws, mechanisms, and secondary causes could
proceed safely, and both medicine and industry be improved, through a
better understanding of God's creation.

As a young man Charles Darwin adopted Paley's vision of natural
history and God completely. He once wrote a friend that he did not
think he "hardly ever admired a book more than Paley's *Natural Theology*:
I could almost formerly have said it by heart." This fact will be key to
making sense of the style, structure, and argument of *On the Origin of
Species*. But we get ahead of our story. For now, a few additional points
should be made. Note that John Ray, William Paley, and company argued
that the study of nature was a pious, Christian endeavor. In doing so they
justified an enormous amount of natural history (including the creation
of one of the largest natural history museums in the world, the Natural
History Museum in London). Combined with Francis Bacon's emphasis
on the divinely approved "discovery" of new plants and animals for commercial and medicinal use, this tradition inspired government support
for the collection of animals and plants from the farthest reaches of a
growing empire.[44]

The connection between natural history, natural theology, and imperialism did more than allow naturalists like Charles Darwin to travel
the world. During the first half of the nineteenth century, the British
cited natural theology's rational demonstrations of God's benevolent
providence to claim divine approval for both imperial expansion and

the industrial-capitalist system driving colonialism. Both Catholic and Protestant missionaries embraced natural history as a means of proving Christianity true. Meanwhile, at home, the British governing classes used natural theology as a rational adjunct to Scripture to defend their power within a highly hierarchical society.[45]

We will trace the outcome of these ties over the course of the nineteenth century, but for now it is important to note the connections that British naturalists, ministers, and physicians were drawing between the study of nature, demonstrations of God's governance, and particular political and economic systems. As we have seen, belief in God's providence did not necessarily equate to doing nothing in the face of suffering. A physician must still try and ameliorate sickness via the study of how God had chosen to govern creation. Paley believed that, properly understood, Christianity supported certain kinds of political, social, and moral reform. Indeed, he argued that the fact God designed a creation in which happiness prevailed over suffering contained clear lessons for human conduct: Christians had a sacred duty to promote the happiness of others. By this measure, Paley insisted that the sins of the British were great, and the potential for reform large. When writing of the "abominable tyranny" of slavery, he demanded that readers reflect on whether the British government, which "had so long lent its assistance to the support of an institution replete with human misery," was "fit to be trusted with an empire the most extensive that ever obtained in any age or quarter of the world."[46]

But how best to reform the world, at what pace, and for whom? William Paley was writing for generations wrestling with the fates of three extraordinary revolutions aimed at ameliorating the suffering of some (depending on the revolution): the American Revolution (1775–1784), the French Revolution (1789–1799), and the Haitian Revolution (1791–1804). The Industrial Revolution was occurring at full speed, and imperial and colonial expansion was transforming landscapes and societies, spreading prosperity for some and misery for others. Change, reform, progress, and even revolution were the slogans of the day. But whose vision of progress should prevail? At what rate should change proceed? And whose children should benefit? Given consensus that knowledge about God and nature went hand in hand, anyone claiming they had an answer to these questions had to either get *both* of God's books on their side or else have a pretty good argument for why one or the other book might be dispensed with in explaining the world.

CHAPTER TWO

From Nature's Coffins to Her Cradles

ON JANUARY 6, 1781, A PROSPEROUS PHYSICIAN NAMED ERASMUS DAR-
win (Charles Darwin's grandfather) took a moment to compose a brief
note to his good friend James Watt (of Watt's steam engine fame) to
explain why he could not attend a Lunar Society meeting. As a collection
of friends interested in natural philosophy, the group called themselves
the Lunar Society because meetings were timed with the full moon to
make travel in the evening less perilous. The society was one of Erasmus's
great delights. But on that cold night in January, Erasmus excused himself on the grounds that "The 'Devil has played me a slippery trick,' sending the measles with peripneumony [inflammation of the lungs] amongst
nine beautiful children of Lord Paget's." He then added:

> For I must suppose it is the work of the Devil? Surely the Lord could
> never think of amusing himself by setting nine innocent little animals
> to cough their hearts up? Pray ask your learned society if this partial
> evil contributes to any public good?—if this pain is necessary to estab-
> lish the subordination of different links in the chain of animation? If
> one was to be weaker and less perfect than another; must he therefore
> have pain as a part of his portion? Pray enquire of your Philosophs,
> and rescue me from Manichaeism.[1]

The letter was sarcastic, but the question was dead serious. Erasmus
Darwin did not believe in the close personal governance of the all-wise,
all-good, all-powerful God of Anglican Christianity. As we have seen,

William Paley firmly believed, as did the founders of the new sciences, that the study of the purposeful parts of animals and plants would prevent science from undermining belief in God's close personal governance over his creation. Erasmus Darwin, however, had no interest in defending Christianity's God. He rebelled from the idea that a benevolent God would "amuse himself by setting nine innocent little animals to cough their hearts up." (The Manichaeism from which he begged to be rescued was the heresy of believing evil, or the Devil, had as much power as God.) Unwilling to believe in an attentive God who would permit such things, Erasmus moved God to so great a distance that he was accused of atheism. And he used an evolutionary concept of species to do so.

We have seen how William Paley and company linked a particular explanation of the origin of species (the "theory of independent acts of creation") to the argument that God is all-wise, all-powerful, and all-good. Both claims were supposed to explain the world, from the hierarchical structure of society (with its unequal distribution of wealth and resources), to the origin and meaning of evil and suffering, including the diseases and deaths of children. Both also reflected a particular explanation of human nature. According to Scripture, humanity was special and unique among God's creatures: created in God's image, they were closest to the angels in their enormous capacity for good; yet as fallen creatures characterized by the sinful propensity to be selfish and do evil, they also shared the nature of apes. (Few claimed human beings were genealogically related to apes rather than Adam. The theory of independent acts of creation, the human capacity for speech, and prevalent interpretations of Genesis precluded such a radical move.) All of these narratives mapped well onto the Thirty-Nine Articles of the Anglican Church.

Erasmus Darwin decided that neither Anglican nor any other kind of Christianity provided the best means of making sense of the world, much less improving it. He also found a very different message in the new sciences. Indeed, the historian John Hedley Brooke notes Erasmus's beliefs as evidence of the fact that "The philosophy of nature that, during the seventeenth century, was upheld as the most protective of the sense of the sacred in nature was the very one that, in later social contexts, was most easily reinterpreted to support a subversive and secular creed."

William Paley was trying to keep the Book of Nature firmly on the side of traditional Christian defenses of belief. Erasmus Darwin, by contrast, was driving the realm of secondary causes into the origin of species, and in doing so imagining a very different way of explaining the state of the world.[2]

INNOCENT LITTLE ANIMALS

Early on in life, Erasmus Darwin began to question whether the deaths of his friends, family members, and patients were due to the will of a close personal God. In 1754, at the age of twenty-three, one paragraph after informing his friend Thomas Okes that his father had departed "this sinful world," Erasmus confessed that he had abandoned belief in the close personal God of Christianity:

> That there exists a superior ENS ENTIUM which formed these wonderful creatures, is a mathematical Demonstration. That HE influences things by a particular Providence is not so evident. The Probability, according to my notion, is against it, since general laws seem sufficient for that end. Shall we say no particular providence is necessary to roll this Planet round the Sun, and yet affirm it necessary in turning up cinque and quatorze, while shaking a box of dies? Or giving each his daily bread?[3]

(Ens Entium means "Being of Beings," and was often used by those not willing to commit to the word God.)

To make sense of Erasmus's statement, it is important to know that theologians had divided God's governance into various kinds of providence: General providence represented the natural laws (like the universal law of gravitation) through which God generally governed; miracles represented breaks in nature's laws for some specific purpose; and particular (or special) providence occurred in between these realms (and thus was often confused with both). Particular providence represented divine action working through natural laws to achieve a specific purpose; say, the use of natural events like the spread of disease to achieve a particular providential outcome. Most Protestants emphasized general and particular providence and relegated miracles to the time of Christ and the Apostles. (They did not want to seem too Catholic by believing in saints interceding on humanity's behalf and miraculous healings.)

But Erasmus demanded to know: Do we imagine that God is in control of how dice fall when playing a game of chance? The question, he assumed, must be answered in the negative. So, too, with the provision of one's daily bread. Erasmus's God did not intervene in human lives. Erasmus's God did not answer prayers. Although Erasmus signed his letters *annuente D.O.M.* ("The most good and most great God nodding"), the concept of particular providence had no meaning for him. A distant, law-governing God was acceptable, but he had no interest in locating a special purpose and meaning in suffering. Good and useful individuals simply should not die young. He replaced God with gods, goddesses, nature, deities, first cause, or *Ens Entium*. In the case of the fatal illness of Josiah Wedgwood's eight-year-old daughter Mary Anne, for example, he spoke of "Nature (or the above-mentioned Devil, as you like)" as the cause that the doctor must route from its path.[4]

Erasmus's deism (belief in a distant God who ruled solely by natural laws) had profound implications for belief in a hereafter. In contrast to John Ray and William Paley, Erasmus had little interest in arguing that the "Light of Nature" afforded arguments for the existence of heaven. He wrote that the fact God had made us out of nothing was the only reason for hope, for surely such a God can "re-create us." Erasmus had *hope*, but this was a far cry from the confident *faith* of John Ray and William Paley.[5]

Although he constantly invoked a first cause in his writings, Erasmus's enemies did not think he believed in God at all. But in an unpublished poem (said to have been written in 1799) entitled *The Folly of Atheism*, Erasmus posed, to all appearances quite sincerely, the question: "Dull Atheist! Could a giddy dance of atoms lawless hurl'd, Construct so wonderful, so wise, So harmonize'd a world?" It was a brief poem, just fifty-six lines. He asked how the atheist would explain the unknown power that "bids the babe to catch the breeze" and expand its "panting breast" to breathe and "with impatient hands untaught, The milky rill arrest." How would they explain the "unextinguish'd love" of the mother who bore the babe "along the rugged paths of life. . . . A God! a God! the wide earth shouts! A God! the heavens reply. He mounded in his palm the world, And hung it in the sky." His power was "great and good."[6]

At this time, individuals willing to press critiques of Christianity into a complete atheism (that there is no God, Christian or otherwise) were rare. After all, "Natural laws themselves presupposed a divine Lawgiver."[7] The Scottish natural philosopher David Hume had asked his damning questions: Who designed the Designer? Why did the Designer have to be Good, when all experience seemed to indicate otherwise? The French satirist Voltaire had ridiculed natural theology's apparent conclusion that "this is the best of all possible worlds." But Hume seemed to think a declaration of "no God" just as illogical, while Voltaire insisted that Julien Offray de La Mettrie—who defended atheism in a 1743 book entitled *Man a Machine*—must be *fou* (crazy).[8] Indeed, historian Ashley Marshall has argued that Erasmus was not trying to "disenchant" the world, but rather trying, amid David Hume's critiques of the design argument, to keep the world sacred by maintaining a place for a benevolent, albeit distant, Creator.[9]

In distancing the Creator from the joys and miseries of creation, and thus abandoning Christianity, Erasmus gained two things: the ability to repudiate belief in hell and, in his mind, open up a wider compass for human agency. The first of these, the doctrine of eternal damnation, had clearly gnawed at the consciences of other individuals as well. A century earlier, John Ray had tried to give reasons for how the eternal damnation of the wicked could be reconciled with the justice and goodness of God. Yet in the end he gave up and noted only that the day of revelation would include an understanding of "what now to our dim-sighted Reason is not penetrable." It was a gentle concession that he could not quite understand why a just, good God would condemn even the most wicked to eternal punishment. Why was Ray unwilling to abandon belief in eternal punishment when he could not understand it? Ray's answer was firm: "If we deny the Eternity of the Torments of Hell, I do not see but that we may, upon as good grounds, with Origin, deny the Eternity of the Joys of Heaven." Give up hell by interpreting Scripture as you like, Ray argued, and heaven must fall as well.[10]

A century later William Paley gave a very practical defense of belief in hell when imagining why God permitted stark social and economic disparities. Faith in a system of rewards and punishments in the hereafter, Paley explained, meant that what matters is not the temporary condition into which we are cast, but our behavior within whatever station to which we were born.

> *Without threat of eternal torture, men want a motive to do their duty, and their rules want authority. Promise them future rewards and a perennial problem is solved; the unequal and promiscuous distribution of power and wealth. The masses will put up with their hardships and degrading stations once they realize that injustice will be rectified in the hereafter. This one truth changes the nature of things. It gives order for confusion, makes the moral world of a piece with the natural.*[11]

This excerpt makes sermons about heaven and hell appear to be entirely about social control. But like Ray, Paley believed that parishioners' souls were at stake. Whether a man performed his duty in the station in which he was born, for Paley, determined both the state of the earth and his eternal salvation in the hereafter. The "truth" was a divine truth and gave "order for confusion" because, according to Scripture and nature, God was wise, good, and just. However, by the end of the century critics of the status quo (like Erasmus) noticed how convenient promises of reward and punishment in the hereafter had become an argument against reform, whether Paley and his fellow ministers were earnest about saving souls or not.

Some were trying to save Christianity by abandoning doctrines that seemed to justify suffering, whether on Earth or in the hereafter. Indeed, some of Erasmus's closest friends were insisting that belief in hell was a corruption of true Christianity. His fellow Lunar Society member, the chemist and theologian Joseph Priestley, denied the Anglican belief in hell on the grounds that it was based on an irrational and unmerciful concept of God.

Priestley insisted on other revisions as well. He denied the doctrine of the Trinity, which in both Catholic and most Protestant doctrines held that God was *both* one and triune (the Father, the Son, and the Holy Ghost). Intent on ensuring that Christianity was also rational, Priestley insisted: "Three divine persons constituting one God is, strictly speaking, an absurdity, or contradiction." Indeed, while he believed that Jesus played an important, divinely appointed role within the faith, Priestley even denied both Christ's divinity and the miracle of the virgin birth. As a Unitarian, Priestley believed that God was one, not three.[12]

These debates over how much Christianity should change to adjust to new standards of both ethics and reason had profound implications for the options available to explain a world in which children died. Take two sermons, one by the Anglican minister Doddridge and the other by the American Unitarian William Peabody, on the loss of children. Both sermons begin with a story in the Old Testament (Kings) in which a Hebrew mother whose child died in her arms is asked, "Is it well with the child?" The mother replies: "It is well." Both ministers interpreted the mother's reply as a praiseworthy recognition that first, such great affliction arose from a wise, powerful, and good providence, and second, that the child is now in heaven. Both also cited the words of Job: "The Lord gave and the Lord taketh away."

Both agreed, as Doddridge reminded his listeners, that "*His kingdom ruleth over all*, and there is *not* so much as *a Sparrow that falls to the Ground without our Father, but the very Hairs of our Head are all number'd* by him." Both Unitarians and Anglicans could take comfort in Doddridge's query: "Can we imagine that our dear Children fall into their Graves without his Notice or Interposition?" Both believed the world could be described in terms of secondary causes, but that ultimately the first cause governed all: "Diseases and Accidents are but second Causes," urged Doddridge, "which owe all their Operations to the continued Energy of the great original Cause."[13] Both agreed all would be revealed in heaven and that all would praise God that, as Peabody wrote, "in those days of earthly sorrow when you half doubted his kindness, His will and not yours, was done." "Children are taken for the parents' sake," Peabody explained, "in order that the parents may have their thoughts carried gently, but irresistibly upward to the heaven of the blessed; the place where they who perish in their innocence, and they whose labor of life is well done, shall be happy forever." (Peabody was minister of the Third Society in Springfield, Massachusetts, and delivered his sermon during a cholera epidemic.)[14]

Peabody and Doddridge differed profoundly, however, on the potential fates of children who died. Unitarians set aside belief in hell on the grounds that no good God would damn even the most wicked for eternity. Doddridge, by contrast, gave sermons advising the pious parents of wicked children because he knew they would rightly fear their children were in a state of "everlasting ruin" and "cast into burnings." Unitarians emphasized God's goodness and mercy; Doddridge left more room for God's vengeance and justice. But clearly Christianity was diversifying, and as a result one might rebel from Anglican doctrines yet not cross Erasmus Darwin's precipice into deism.[15]

It is important to note that one of these positions (the Anglican one) was sanctioned by the British state; the other was not. This is just one reason why Unitarians, as dissenters excluded from political power, tended to be reformers and radicals. At the beginning of the nineteenth century, choosing between Anglican Christianity and a dissenting faith like Unitarianism came with profound political and social implications. In dissenting from one or more of the Thirty-Nine Articles of the Anglican faith, Unitarian ministers could not perform the sacraments of either marriage or burials. Non-Anglicans of all kinds (Jews, Unitarians, Methodists, Baptists, and much less atheists) could not attend Cambridge or Oxford or serve in Parliament.

Meanwhile, Anglican ministers had a response to those who judged God according to human standards. "And what shall we say?" asked Reverend Doddridge. "Are not the Administrations of his Providence wise and good? Can we *teach him Knowledge*? Can we tax him with Injustice? Shall the Most High God learn of us how to govern the World, and be instructed by our Wisdom when to remove his Creatures from one State of Being to another?"[16] Erasmus Darwin confidently replied: "*Yes*, I do tax the Anglicans' God with Injustice." As for Unitarianism, Erasmus called it "a featherbed to catch a falling Christian." In contrast to his friends Joseph Priestley and Josiah Wedgwood (Charles Darwin's maternal grandfather), Erasmus gave up on Christianity, including all God's close personal governance over his creation. He repudiated the idea that God was present amid anything a physician might meet at the bedside of the afflicted, especially the children—those "innocent little animals"—of his friends and patients.[17]

Having abandoned Paley's God, Erasmus had to then confront the following problem: In what sense did affliction have purpose and meaning once disease and death were no longer divinely sent reminders of either man's depravity or the prospect of salvation? He found his answer in the Royal Society's argument that a better knowledge of nature's laws would provide the means of ameliorating suffering in future. In other words, he found his answer in a firm belief in progress via science even as he set aside some of the fundamental premises of the society's founders. As we have seen, emphasizing natural laws as God's means of governance *within certain boundaries* was originally neither radical nor heretical. The early members of the Royal Society, from Isaac Newton to John Ray, were devout Christians who viewed the discovery and description of nature and natural laws as a pious means of learning about God through the "Book of Nature." They had even expressed confidence that the assumption that natural laws and secondary causes are *uniform* (i.e., generally not broken by miraculous intervention) was the only basis by which medicine could be improved. But Boyle, Ray, and later William Paley were confident that because species and purposeful parts could only be explained via independent design, science's emphasis on secondary causes could not remove God's close personal control—his "special providence"—from the joys and miseries of human experience. Thus, the purpose and meaning presumed to exist in all affliction would be secure. Erasmus, by contrast, embraced the study of "general laws" with *no exceptions*, as the only means of both assuring human progress and upholding God's benevolence. In other words, he abandoned the idea that species were independently created by God at the beginning of time.

Erasmus framed his strict commitment to explaining all via secondary causes in the language of piety and belief in God's benevolence by explicitly linking nature's uniformity to the promise of progress. He was not alone. A few months after ridiculing explanations of why God allowed innocent children to cough up their insides, he had read the latest version of Doctor Balguy's *Divine Benevolence Asserted; And Vindicated from the Objections of Ancient and Modern Sceptics*. Balguy argued that it was more probable that "God is *good* rather than *capricious*; because the course of nature is *uniform*. Whatever events befall us, good or bad, arise from certain *general* principles in the constitution and government of the universe." Balguy then added: "We may observe, that, as knowledge increases in the world, the number of *remedies* increases, both against outward pains and sickness; and that the pains men suffer, make them more cautious, and more attentive to discover *new* remedies." Erasmus wrote to his friend Thomas Day that he thought Balguy's book "a very able performance, as his sentiments coincide with my own. He makes it evident that everything was made by the Lord, with design of producing happiness; and seems to show there is more happiness than the contrary in the world." This is a clear confession of belief in a benevolent Creator, but it was not the close personal God to whom one might pray and expect an answer. Particular providence was gone. But Erasmus Darwin clearly saw an advantage in shifting responsibility for great affliction *away* from God, namely the provision of a great deal of work to do, for himself and his friends, in discerning those "general principles in the constitution and government of the universe" and ameliorating the suffering that arose from humanity's tremendous ignorance of nature's laws. [18]

THE ONLY RECOURSE WHICH A PHILOSOPHER CAN FLY TO

Driven by the things he thought wrong in the world, Erasmus found work to do everywhere. Just traveling on the roads of England, whether one was a prosperous physician like Erasmus Darwin or not, was a perilous endeavor. Especially at the beginning of the eighteenth century, the roads were notoriously bad (they were "ribbon dungheaps," in one historian's choice phrase), deeply rutted by wagon wheels and livestock. Accidents and highwaymen threatened life and limb, even as hints of improvements appeared. By 1770, 15,000 miles had been turn-piked (turned into toll roads), which meant more inns and coach services, more freight moving faster and safer, the beginnings of the best postal service in the world, and, unfortunately, more highwaymen. Along with canals, these changes formed the infrastructure upon which the extraordinary transformations associated with the Industrial Revolution soon moved. The slowly improving roads were also the means by which, beginning in 1765, Erasmus Darwin traveled to meetings of the Lunar Society.[19]

Dr. Darwin felt every bump on those roads on his behind. In lamenting that he would once more miss a Lunar Society meeting, fighting the "perpetual war" waged against doctors by the "infernal Divinities" who visited mankind with diseases, he wrote of himself as "imprisoned in a post chaise, . . . joggled and jostled, and bumped and bruised along the King's high road, to make war upon a pox or a fever!" And then he tried to figure out how to alleviate the suffering of which he complained. To his friend Albert Reimarus he wrote that "to hang a coach so easy that a person may read in it a small Print, would be of great use and pleasure." He sketched out some potential designs and sent those as well.[20]

Erasmus put his money toward the solutions to the problems he noticed, funding many of the new canals and turnpikes, for example, to improve transportation. His letters are filled with problems followed by an abrupt transition to "But what then is to be done?" Often these problems were medical, as befit the daily queries that arrived from patients. He supported smallpox vaccination, writing to Edward Jenner of his hopes "that in a little time it may occur that the christening and the vaccination of children may always be performed on the same day." But the question "what then is to be done" was also asked about inefficient farming methods and how to take advantage of new modes of transportation to move goods.[21]

Having lost belief in the power of prayer and the surety of heaven, both Erasmus and his friend James Watt responded to tragedy and grief by keeping busy. In the wake of child loss, Watt insisted the best consolation was to turn the mind to another subject. After his daughter Jessie's death, Watt chose the study of medicinal airs, and he and Erasmus wrote back and forth about apparatuses for delivering various "airs" for treating asthma and other ailments. When Watt lost his second daughter Margaret when she was twenty-nine, Erasmus agreed that "activity of mind is the only circumstance which can prevent one from thinking over disagreeable events." (He had Margaret's death and the war with France both in mind.) "Activity does not always produce pleasure," he confessed, "but I think it always prevents or lessons present pain; and is therefore perhaps the only recourse which a philosopher can fly to in the hour of affliction."[22]

In abandoning God's close personal providence, divine meaning in affliction, and the surety of heaven, Erasmus placed both the cause of and the ability to alleviate suffering entirely in the hands of humanity. He heightened the burden on physicians, but held out a new kind of immortality as a possible reward of "doing good." In *The Botanic Garden*'s ode to the liberating power of reason to vanquish the prison of superstition, Darwin wrote the following of Benjamin Franklin's work on lightning:

> Immortal FRANKLIN sought the fiery bed;
> Where, nursed in night, incumbent Tempest shrouds
> The seeds of Thunder in circumfluent clouds,
> Besieged with iron points in his airy cell,
> And pierced the monster slumbering in the shell.[23]

For Erasmus, the only certain immortality was gained through making new discoveries of aid to humanity. Indeed, he suspected the fact children of the future might know one's name was all the "eternal life" one could or should hope for.

There were additional reasons to work for others, Erasmus insisted. Long experience with his patients convinced him that "one must do something," or else one grows weary of life. Those who "left off business," he wrote, tended to become drunkards or hypochondriacs. And since, Erasmus reasoned, one must do something, "one may as well do something advantageous to one's self or friends, or to mankind." Erasmus's letters are full of ideas for "improvements," which were destined (like the steam engine) to transform the landscape, industry, and society, not just in Britain, but around the world.[24]

Meanwhile, Erasmus had absorbed new ways of thinking about medicine, nature, and God during his time as a medical student in Edinburgh, Scotland. Unlike Cambridge and Oxford, the University of Edinburgh did not require a religious test (a formal affirmation of belief in the Thirty-Nine Articles of the Church of England), which meant the town was full of often quite radical dissenters. Within this world of new ideas and ways of thinking, Darwin learned physician William Cullen's theory that the nervous system (and not some balance of the four humors, as the ancients held) was the seat of physiological balance. This new theory of the body led to new theories of disease (although the treatments tended to remain the same). Cullen held, for example, that "all pathology originated in a disordered action of the nervous system."[25]

For Erasmus, the art of relieving pain consisted in reducing nervous tension via bleeding and purging (strengthening tonics would be used if tension was too low). One must first evacuate the patient via moderate bleeding, and then prescribe vomiting and purging (using senna, mercury, jalap, and other purgatives) on alternate days for a week. This process prepared the body to receive and take advantage of the prescribed medicine, such as bark (quinine), opium, and sometimes steel. (Given the ubiquity of intestinal worms, purging could have beneficent effects, and the iron in the steel could alleviate anemia.)[26]

In some cases, it was just as well that Erasmus's ideas for new treatments did not come to fruition. When Watt's daughter, Jessie, was dying of consumption (tuberculosis), Erasmus urged him to rig up a mechanism by which she could "swing horizontally for 1/2 an hour or longer 4 or 6 times a day till she becomes vertiginous and sickish." The idea was that by acting as a nonchemical emetic, the swinging would heal the ulcers in her lungs. Surely such a contrivance, he wrote, would be preferable to all other modes of inducing the patient to vomit. (He was right—the other modes generally caused violent, repetitive retching.) In response, Watt did draft a design for a "movable rotative couch" for human centrifugation, though he refused, apparently, to subject his daughter to the new treatment.[27]

Erasmus's most famous book, *Zoonomia, or the Laws of Organic Life* (1794), was the result of a lifetime's effort to apply the new sciences and a purely mechanistic view of nature to medicine. The title page declared that the book contained "the immediate causes of animal motions, deduced from their more simple or frequent appearances in health, and applied to explain their more intricate or uncommon occurrences in diseases." Volume II, published in 1796, contained a classification of diseases (this was the age of Linnaeus, and Erasmus Darwin was one of the Swedish taxonomist's biggest fans) into orders, genera, and species, including "their methods of cure." Printed in three volumes, *Zoonomia* was a 1,125-page reflection of Darwin's belief that man could *do* better in the face of illness, disease, and even death, if only he *understood* nature better.

Although (as we will see in the next section) *Zoonomia* contained a theory of evolution, it was Erasmus's efforts at reforming medicine that inspired the most comment and (in contrast to his evolutionary views) almost universal praise. "Among modern efforts to diminish the calamities of the world by improving the art of medicine," wrote the *Analytical Review*, "those of the present author will . . . as far as we can anticipate the decision of time be crowned with the amplest success." Another wrote that *Zoonomia* "bids fair to do for medicine what Sir Isaac Newton's *Principia* has done for Natural Philosophy." This was high praise indeed.[28]

Erasmus's efforts to improve medicine were driven by constant reminders of physicians' helplessness in the face of suffering. He lost his beloved wife Polly to "a very painful Disease, of many Months" (probably gallstones) in 1770. The impact of this loss on his days and patterns of thought was profound: he set all metaphysics, which he obviously loved, aside for a decade. Their son Charles, a promising young medical student, died at the age of nineteen in 1778. With no reason to wear gloves (germ theory was almost a century away), Charles had cut his finger while dissecting the brain of a child who had died of "hydrocephalus interns" and soon became delirious. A good physician in these days could do little to cure somatic disease but was an expert in prognosis. And Erasmus was a good physician. Upon arrival at his son's bedside, he accurately predicted "this valuable young man" would be dead in three days.[29]

Two of the five children borne by Polly, William and Elizabeth, died in infancy. Later in life, Erasmus had two daughters with his mistress Mary Parker, and seven children with his second wife, Elizabeth Collier (one of whom, Henry, died at the age of one). His son Erasmus killed himself in the throes of what was called a "morbid depression." Another son, Robert, would live longer: he would marry Susannah Wedgwood and have six children, including a son born in 1809 named Charles Darwin.

And of course there were the ailments of his patients, including close friends and their children. These included the wasting disease of consumption, which destroyed whole families and cut off the young in the prime of life ("Go on, dear Sir," Erasmus exhorted the author of a treatise on consumption, "Save the young and the fair of the rising generation from premature death; and rescue the science of medicine from its greatest opprobrium"); whooping cough, so violent that a child's eyes could become inflamed (Erasmus recommended bleeding); and the death of mothers from childbirth (one could not be safe for some time after delivery, and Erasmus pondered why the illness often set in on the eleventh day, but never found the answer). In wrestling with the proper age for smallpox inoculation (he was a firm proponent of both inoculation and, after 1796, vaccination), he reported that according to the bills of mortality of London, Paris, and Vienna, 350 children out of every 1,000 died under two years old.[30]

Wrestling with all this suffering, Erasmus Darwin has been accused of getting his patients addicted to opium. Certainly, his poor wife Polly could not bear, in her final years, to be without it. But faced with the agonizing pain of gallstones, kidney stones, and a whole host of other ailments that drove human beings to distraction, the physician had nothing else to offer. Some of Darwin's treatments (like those of most eighteenth-century physicians) have been ridiculed, for they included bloodletting and purgatives like calomel (a mercury chloride) and tartar emetic. But based on his letters to friends, family, and patients, Erasmus clearly preferred mild remedies, including temperance, exercise, and occupation. He also understood the role imagination and attention could play in the production and curing of some ailments. Historian Patricia Fara notes that, in an age before antibiotics, when a physician could do very little except help a sufferer die comfortably, "Darwin was popular with his patients because he offered them great psychological support as well as liberally prescribing opium to numb their pain."[31]

Erasmus was confident that medicine could be improved via disciplined methods. Unlike some of his contemporaries, he had a broad view of who should benefit from hoped-for improvements in the moral and material condition of humanity. At a time when some of his colleagues were using the Bible, the new sciences, or both to classify some humans as *nonhuman* in the interest of justifying slavery, Erasmus stood ready to intervene if anyone tried to kidnap Olaudah Equiano, whose 1789 book about his life as a slave was inspiring the abolitionist movement. Faced with the knowledge that muzzles for human beings were being made in Birmingham, Erasmus proposed that an exhibit of such instruments before the House of Commons be arranged. "An instrument of torture of our own manufacture," he wrote to James Watt, "would have a great effect, I dare say."[32] And he composed poems in defense of the unity of man, most famously in *The Botanic Garden*, which contained the following lines:

> Hear, oh, BRITANNIA! potent Queen of isles,
> On whom fair Art, and meek Religion smiles,
> Now AFRIC'S coasts thy craftier sons invade
> With murder, rapine, theft,—and call it Trade!
> The SLAVE, in chains, on supplicating knee,
> Spreads his wide arms, and lifts his eyes to Thee;
> With hunger pale, with wounds and toil oppress'd,
> "ARE WE NOT BRETHREN?" sorrow chokes the rest;
> AIR! bear to heaven upon thy azure flood
> Their innocent cries!—EARTH! cover not their blood![33]

Erasmus imagined "The SLAVE, in chains, on supplicating knee" begging for freedom from whites, an image his friend Josiah Wedgwood engraved on pottery to sell to white abolitionists who thus felt they were doing something for the cause. Meanwhile, enslaved men and women in the French colony of Saint Domingue, under the leadership of Toussaint L'Ouverture, were soon wresting freedom for themselves from the most powerful army on Earth. Their actions would become known as the first successful slave revolt in history and lead to the creation of the second independent republic in the hemisphere, the nation of Haiti, on January 1, 1804.[34]

Abolitionists would cite the Haitian Revolution as evidence that enslaved people both hungered for freedom and were capable of self-government. But soon both the Book of Nature and Book of Scripture would be harnessed to stop such ideas from spreading. Slavery's defenders cited the presence of slavery in the Bible and the fact that nowhere did Jesus explicitly condemn the practice. Within a generation, U.S. Senator James Barbour drew upon natural theology's "greater good" arguments to insist, in 1820, that "However dark and inscrutable may be the ways of heaven, who is he that arrogantly presumes to arraign them?" Slavery, Barbour argued, was "a link in that great concatenation which is permitted by omnipotent power and goodness that must issue in universal good."[35]

Abolitionists countered that surely explicit biblical condemnation of slavery was not needed given Christ's messages of "Love Thy Neighbor" and the Golden Rule ("Therefore all things whatsoever ye would that men should do to you: do ye even so to them"). They, too, spoke much of God's benevolence but found a very different message in nature's design. When, in 1744, Virginia-born physician John Mitchell published a mechanistic explanation of the origin of different skin colors (based on the refraction of light upon different densities of skin) in *The Philosophical Transactions of the Royal Society of London*, he declared that difference to be indicative of God's benevolence in rendering African skin "more insensible" to a hot, humid climate. Abolitionists Olaudah Equiano and David Rittenhouse cited Mitchell's work as "scientific" evidence that "the sin of slavery" persisted "merely because *their* bodies may be disposed to reflect or absorb the rays of light, in a different way from *ours*." Apologists for slavery, however, quickly turned Mitchell's work into the claim that both Africans and their descendants possessed a natural, divinely designed physical hardiness that made them immune to external injuries. They then used this mechanistic explanation of skin color to declare slavery (and its violence) both natural and divinely approved. (As we will see, moves like this would be just one reason why, although abolitionists like Frederick Douglass would decide that belief in special providence must go for slavery to end, they remained wary of accepting nature's laws as a helpful alternative.)[36]

Most abolitionists countered the use of physiology and anatomy to draw distinctions between human beings with passages from the Bible. They cited Acts 17:26, for example, as scriptural evidence that God "hath made of one blood all nations of men for to dwell on all the face of the earth." But clearly Erasmus Darwin was uninterested in calling upon Scripture to explain the world, much less change it. So instead he turned to God's other book, the Book of Nature, certain that progress, science, and reason must be on the side of justice for everyone. As we will see, that is not exactly how things turned out. Erasmus does not seem to have realized that Nature might be just as fallible a guide to justice as he believed the Bible was.

Would It Be Too Bold to Imagine?

Erasmus's vision of what nature is like was strongly rooted in his commitment to the idea that society could change for the better via democratic reform, if not revolution. When the French Revolution broke out in July 1789, there were seven children under the age of nine in Darwin's home. Erasmus watched events with an eye to the world being built for future generations: "Do you not congratulate your grand-children on the dawn of universal liberty?" he exclaimed in a letter to Watt. "I feel myself becoming all french both in chemistry and politics." Meanwhile, he worked away at *Zoonomia*. There, Erasmus rebelled against the tight links that had been forged between belief in a personal God and the independent creation of species by proposing that species arose via purely natural processes from other species (in other words, species evolved). William Paley had defended the independent creation of species as the final citadel of a rational faith in God's protective providence, carefully documenting how the purposeful parts of animals, including human anatomy, must have been designed by God. Erasmus, by contrast, decided that God's benevolence must be found solely in the fact that species advanced over time. He also explicitly connected his vision of nature to justifications of radical social change to ameliorate human suffering and ensure the world would be better for all children.[37]

When *Zoonomia* finally appeared in 1794 (one year after the revolution in France had descended into the chaos of the Great Terror), readers found the following passage:

> *Would it be too bold to imagine that, in the great length of time, since the earth began to exist, perhaps millions of ages before the commencement of the history of mankind, would it be too bold to imagine, that all warm-blooded animals have arisen from one living filament, which the Great First Cause endued with animality, with the power of acquiring new parts attended with new propensities, directed by irritations, sensations, volitions and associations; and thus possessing the faculty of continuing to improve by its own inherent activity, and of delivering down these improvements by generation to its posterity, world without end!*[38]

As we saw in chapter 1, John Ray, Robert Boyle, and other founders of the new sciences had studied the purposeful parts of animals and plants (as a rational accompaniment to studying Scripture) in order to show that the God who governed via natural law was still the God who governed the fall of every sparrow. Having given up on a personal God who created species "in kind," Erasmus decided, by contrast, that the world was "generated" rather than "created." Species arose from *other species* via a purely mechanistic, naturalistic process rather than by creative, supernatural acts of God. But how, precisely? Erasmus imagined that animals had acquired their wonderfully purposeful parts—the elongated trunk of the elephant, the strong jaws or talons of predators, the hard beaks of sparrows that could crack seeds—"during many generations by the perpetual endeavour of the creatures to supply the want of food." In other words, animals used some organs more, less, or differently as they tried to survive, and any consequent changes in structures were "delivered to their posterity with constant improvement of them for the purposes required." (This kind of mechanism is called the inheritance of acquired characteristics, as it assumes that individual variations produced by either environmental conditions or increased use or disuse will be passed down to the next generation.)[39]

Erasmus thus declared that the entire living world arose from "the activity of its inherent principles" rather than by a sudden Almighty fiat. Species arose from other species; life arose via spontaneous generation from nothing via combinations of matter and electricity. To Erasmus, God had provided the means for creation to create, enrich, and improve itself. He believed such a view reflected a grander vision of God: "What a magnificent idea of the infinite power of the GREAT ARCHITECT!" he rejoiced, "THE CAUSE OF CAUSES! PARENT OF PARENTS! ENS ENTIUM!"[40]

Imagining species could arise from other species solved several problems for Erasmus. The theory solved, for example, mysteries like why some parts of animals apparently served no purpose. The gradual formation and improvement of the animal world explained "the breasts and teats of all male quadrupeds, to which no use can be now assigned" as arising from their hermaphroditic past. Perhaps more importantly, Erasmus's visions of nature reconciled belief in God's benevolence with the tremendous amount of suffering in the world and explained why anyone should do good, both without appealing to Scripture. For example, he described his theory that struggles in the present formed the seeds of improvement in the future as perfectly "consonant to the idea of our present situation being a state of probation, which by our exertions we may improve, and are consequently responsible for our actions."[41]

Historian Norton Garfinkle has described how, for the most part, reviewers of *Zoonomia* ignored the book's evolutionary ideas. The ideas that life could be spontaneously generated or species arise from other species were considered too speculative to deserve much attention. But nearly a decade later Erasmus tried again, this time in the form of a long epic poem entitled *The Temple of Nature; Or, The Origin of Society: A Poem, with Philosophical Notes*. Like *Zoonomia*, the poem is often cited in the history of evolution theory as a poetic precursor to Erasmus's more famous grandson's work. The poem, published in 1803, was a long and elaborate imaginary account of the origin of the cosmos, life, and human society.[42] The most famous lines are those describing the first appearance of organic life:

Hence without parent by spontaneous birth
Rise the first specks of animated earth;
From Nature's womb the plant or insect swims,
And buds or breathes, with microscopic limbs.
ORGANIC LIFE beneath the shoreless waves
Was born and nurs'd in Ocean's pearly caves
First forms minute, unseen by spheric glass,
Move on the mud, or pierce the watery mass;
These, as successive generations bloom,
New powers acquire, and larger limbs assume;
Whence countless groups of vegetation spring,
And breathing realms of fin, and feet, and wing.[43]

Notice Erasmus is imagining organisms acquiring new powers and diversifying into fish, reptiles, mammals, and birds, a vision of triumphant self-creation via spontaneous generation and vague mechanisms like "propensities" and "appetencies." Most important for our purposes is the fact that the poem also contained an alternative explanation of the existence of suffering. First, Erasmus listed all the terrible things one might observe in nature. He described the great slaughterhouse of the warring world: predation, pestilence, famine, earthquakes, and flood.

Air, earth, and ocean, to astonish'd day
One scene of blood, one mighty tomb display!
From Hunger's arm the shafts of Death are hurl'd
And one great Slaughter-house the warring world![44]

The poem is full of footnotes explaining the scientific evidence for his vision of the world. For this passage, the footnote explained how "The stronger locomotive animals devour the weaker ones without mercy. Such is the condition of organic nature! whose first law might be expressed in the words, 'Eat or be eaten!' and which would seem to be one great slaughter-house, one universal scene of rapacity and injustice!" Looking for evidence of God's goodness in nature, Robert Boyle had written of the goodness of Providence in bestowing upon insects instinctive knowledge of the best place to lay their eggs. But Erasmus reminded the reader of a gadfly burying "her countless brood" in the bodies of cows and horses, from whence the "hungry larva eats its living way, Hatch'd by the warmth, and issues into day." A corresponding footnote further explained how, "adhering to the anus (the fly) artfully introduces itself into the intestines of horses, and becomes so numerous in their stomachs, as sometimes to destroy them; it climbs into the nostrils of sheep and calves, and producing a nest of young in a transparent hydatide (cysts) in the frontal sinus, occasions the vertigo or turn of those animals." He cited Linnaeus as an authority for the claim that in Lapland these flies attacked reindeer in such great numbers that herders had to leave the woods and go to the mountains.[45] The next lines described how

> The wing'd Ichneumon for her embryon young
> Gores with sharp horn the caterpillar throng.
> The cruel larva mines its silky course,
> And tears the vitals of its fostering nurse.

Here is the footnote attached to the above lines:

The wing'd Ichneumon: Linnaeus describes seventy-seven species of the ichneumon fly, some of which have a sting as long and some twice as long as their bodies. Many of them insert their eggs into various caterpillars, which when they are hatched seem for a time to prey on the reservoir of silk in the backs of those animals designed for their own use to spin a cord to support them, or a bag to contain them, while they change from their larva form to a butterfly; as I have seen in above fifty cabbage-caterpillars. The ichneumon larva then makes its way out of the caterpillar, and spins itself a small cocoon like a silk worm; these cocoons are about the size of a small pin's head, and I have seen about ten of them on each cabbage caterpillar, which soon dies after their exclusion.[46]

Nearly a century later, Charles Darwin would cite the Ichneumon wasps when expressing his doubt regarding the existence of a benevolent, personal, designing creator.[47] But his grandfather Erasmus had raised the problem first. Faced with all this terrible suffering (and he imagined caterpillars did suffer), Erasmus wondered:

> Ah where can Sympathy reflecting find,
> One bright idea to console the mind?
> One ray of light in this terrene abode,
> to prove to Man the Goodness of his GOD?[48]

Erasmus's answer demanded two things of his readers: first, one must judge God's goodness by weighing the amount of good and bad in the world while paying attention to the scale at which one was measuring. Surely, he argued, if one was able to step back and see the entirety of creation, the good would outweigh the bad. Again, a footnote explained: "The sum total of happiness of organized nature is probably increased rather than diminished, when one large old animal dies, and is converted into many thousand young ones; which are produced or supported with their numerous progeny by the same organic matter."[49] One must always keep the long view in mind, and remember that death was always, inevitably, accompanied by new births.

> *When we reflect on the perpetual destruction of organic life, we should also recollect, that it is perpetually renewed in other forms by the same materials, and thus the sum total of happiness of the world continues undiminished; and that a philosopher may thus smile again on turning his eyes from the coffins of nature to her cradles.*[50]

This claim (that the happiness and good in the world outweighs the bad) actually sounds a great deal like passages in William Paley's *Natural Theology*. But Erasmus added a second, radical idea to the claim that the world was, on balance, happy and good: both goodness and happiness, he believed, *increased* over time via a progressive natural process of the transmutation of species into new and improved forms. *The Temple of Nature*'s long ode to the life-giving forces of LOVE and REPRODUCTION (Erasmus italicized or capitalized his favorite ideas) emphasized the constant production of life that accompanied constant death. "*Hail the DEITIES OF SEXUAL LOVE*," he cried, that regenerated the world through the creation of not only new life, but *new forms*. And how, he demanded, could this be so, if a benevolent Creator was not behind it all? The progress of life proved the existence of a first cause, albeit as distant legislator rather than attentive Father. This was true even if it meant that human beings had come from apes, rather than being independently created in the image of God.[51]

The distance to which he was potentially removing God's governance by imagining species creating themselves did not worry Erasmus. He had no trouble seeing the transmutation of one species into another as consonant "to the dignity of the Creator of all things." For him, a vision of nature governed entirely by natural laws provided a better explanation of human experience and natural history, from troubling creatures like the Ichneumon to innocent children coughing up their insides. Nor did he fear dethroning humanity from the privileged place of having been created in God's image. The passage in Matthew reminding the Christian reader of God's close personal providence assumed that humans were worth more than sparrows: "Fear ye not therefore, ye are of more value than many sparrows." Erasmus, by contrast, seemed comforted by the belief that humanity might humbly accept not only swallows but emmets (an old word for ants) as his brothers and worms as his sisters:

> Stoop, selfish Pride! survey thy kindred forms,
> Thy brother Emmets, and thy sister Worms![52]

Erasmus found comfort in this family tree. For if he was right, then human beings must neither expect any better than their fellow creatures, nor despair that they and their children were subject to the same natural laws as all other animals. He also found what, in his eyes, was the surest means of securing humanity and justice for all human beings, and indeed all sentient creatures. If all humanity were brethren, related by common descent, then no grounds existed, he was sure, for declaring one group of human beings superior to another. Of course, Paley would have agreed with the final point on biblical grounds. Erasmus, however, had decided

Paley's brand of Christianity had its chance and had proved too easily harnessed to humanity's sinful, hypocritical ways.

I THINK ABOUT THE LITTLE THING ALL DAY

When *The Temple of Nature* appeared in 1803, the response from those fearful that ideas about a self-governing nature would be used to justify political revolution, or worse, was swift. The *Anti-Jacobin Review* declared the poem full of "monstrous absurdities" and soundly damned it for abandoning the Mosaic story of creation and denying "interference of a Deity in the creation and preservation of everything that exists." Even journals that had praised *Zoonomia* ridiculed the poem as substituting a "religion of nature" for the religion of the Bible.[53]

The political context of the time is key to understanding the passion with which those in power condemned Erasmus's work. By 1803 the French Revolution of 1789 had descended into chaos. During the Great Terror of the summer of 1793, the revolutionaries guillotined the king and queen of France and then began beheading each other. The reaction to the revolution among the ruling classes in Britain, who had the most to lose if the revolution crossed the channel, was swift. New journals were established to emphasize the authority of Scripture and ridicule books like *Zoonomia*. Meanwhile, an evangelical revival emphasized a return to the Bible, rather than nature, as the most important means of knowing God. Erasmus's friend and fellow Lunar Society member, the Unitarian Joseph Priestley, was chased out of the country by a "Church and King" mob who burned down his house and laboratory. (Priestley fled to the United States, where he was welcomed by Thomas Jefferson.)[54]

The *Anti-Jacobin Review*, founded in 1797 to oppose "sedition and irreligion," had Erasmus in its sights from the start. The newspaper's founders connected Darwin's evolutionary speculations to the idea that man could create a new society from the ground up by use of his reason: evolution, in other words, was associated with utopian, democratic visions of the progress and perfectibility of man. For had not the French revolutionaries called on Count Buffon's mad idea that species can change into other species to justify the destruction of God, king, and country? (In the 1750s Buffon had proposed in his encyclopedic survey *Natural*

History: General and Particular that a small number of originally created forms diversified as they moved into different environments.) Fear of how political radicals were using the idea of transmutation was explicit in the response to Erasmus's work. *The Temple of Nature* imagined a democratic-like process of change powered via the efforts of millions of individual beings, rather than design from on high by God or governance from on high by a monarch. What one actually observed in nature, argued critic of the French Revolution Edmund Burke, was subordination, order, stability, and hierarchy, all imposed from above. In this view, both natural history and anatomy provided lessons in the divinely designed goodness of the status quo (i.e., a highly hierarchical society in which one's station in life was assigned by God and obedience was the best route to reward in the hereafter).[55]

It is important to note that when *The British Critic* called Erasmus's theory an insult on good sense, it was a fair point according to contemporary concepts of good science. (Indeed, Erasmus's theory is extraordinarily speculative according to today's standards as well.) Most working naturalists agreed that Erasmus's ideas might provide entertaining fare as poetry but did not count as science. The most renowned naturalist of the era and director of the Museum of Natural History in Paris, Georges Cuvier, criticized the idea that species can change based purely on comparative anatomy: First, each part of a species' anatomy is perfectly correlated to every other part, and therefore change in one part would kill the animal. Second, species vary, but never to the point a naturalist would declare them a new species. Third, both the anatomy and behavior of species are perfectly correlated to their "conditions of existence," which means any change in the organism would render it no longer adapted to its environment.[56]

As a Protestant Christian in Catholic France, Cuvier was no friend of extensive entanglements between theology and science. He grounded his argument against evolution in observation and logic: no one had seen one species change into another, and no rational mechanism could be imagined through which purposeful organs could arise via a purely mechanical process. He also relied on a widely accepted consensus regarding what constitutes scientific knowledge. Origin questions were, to Cuvier, outside the purview of what naturalists and anatomists could study since they happened in the deep past and thus could not be observed. Naturalists must assume creation, and then deal in things that could be observed. (As we will see, Erasmus's grandson would have to deal with this problem, and most French scientists, including Louis Pasteur, would *not* be convinced by his solution.) Meanwhile, in Britain, naturalists united Cuvier's argument with Paley's more theological version to solidify the theory of independent acts of creation as the most pious and rational view of the origin of species.

It did not take long for radicals to put Erasmus's ideas (scientific or no) to political use. They had long been up against arguments that a static nature proved conservative ideologies both Godly and true. But Erasmus's vision of the past offered new explanations of the human condition while imagining very different things for humanity's future. One of the first and most famous reimaginings of the world inspired by his work was composed during the summer of 1816 in a castle on the shores of Lake Geneva, Switzerland. Trapped by bad weather and with nothing to do, a group of friends had been challenged by their host, Lord Byron, to write a ghost story. Eighteen-year-old Mary Wollstonecraft Shelley took up the challenge by imagining a human being, rather than God, creating life through purely material means. She later wrote that she was inspired by her friends' discussion of "various philosophical doctrines . . . among others the nature of the principle of life, and whether there was any probability of its ever being discovered and communicated." Some experiments of Dr. Darwin had been mentioned ("I speak not of what the Doctor really did," Shelley added, "or said that he did, but, as more to my purpose, of what was then spoken of as having been done by him") in which Darwin had "preserved a piece of vermicelli in a glass case, till by some extraordinary means it began to move with voluntary motion." Might life be given after death, she wondered? "Perhaps a corpse would be re-animated; galvanism had given token of such things: perhaps the component parts of a creature might be manufactured, brought together, and endued with vital warmth."[57]

Shelley called her story *Frankenstein: The Modern Prometheus.* (Prometheus was the Greek god who created humans from clay and then gave them the gift of fire, control over the elements, and the promise, or curse, of scientific and technological progress.) As she wrote, famine threatened Europe. The continent was in the throes of a food shortage caused by the upheaval of the Napoleonic Wars, the British Corn Laws (tariffs on imported grain to protect landowners' profits), and temporary climate changes following the eruption of Mount Tambora. In the wake of great suffering among the poor, political revolution threatened, again. At the time, Shelley herself was grieving for her firstborn, a daughter who had lived just eight days before dying a year before. Like so many mothers and fathers, Shelley would lose several more children. Her second child, a boy named William, would die at the age of three of malaria: "The misery of these hours is beyond calculation," Mary wrote during his final illness. "The hopes of my life are bound up in him." By the time she completed *Frankenstein* on May 14, 1817, she was several months pregnant with her third child. This daughter, Clara, would die at the age of one of dysentery.[58]

How to explain all this suffering? For many of Mary Shelley's generation, the appropriate, pious response to suffering, including child loss, was clear: resignation to God's will, supported by faith in the presence of a divine purpose behind every affliction. That did not mean mothers and fathers did not feel soul-shaking grief when their children died. Nor did it mean that they did not try to ameliorate the suffering of children who still lived. One must still call a physician. One must still fight to end slavery and other social ills. But the route to solace amid loss was supposed to be clear: faith in God's goodness and the consequent possibility of reunion in heaven. This faith was in turn rooted in the Genesis story of man created "in His image" with a privileged status, separate from animals, and partaking more of the angels than the apes. All was rooted in the assumption, based on the Book of God's Word and his Works, that not a sparrow falls to the ground without God's attention.

Mary Shelley was raised, however, in a household infamous for embracing radical alternatives to the consensus that both Scripture and nature demonstrated God's close personal control over human lives. Her father, William Godwin, assumed that suffering arose entirely from an unjust social organization that might be, and must be, changed. Her mother, Mary Wollstonecraft, famously applied that assumption to women's suffering in her 1792 book *A Vindication of the Rights of Women*. Mary Shelley's husband Percy Shelley's explanations of the world were even more radical, for unlike Godwin and Wollstonecraft, he dispensed with God altogether. He was expelled from Oxford for writing a pamphlet entitled *The Necessity of Atheism*. (Shelley argued God was an unprovable "hypothesis" that concealed humanity's ignorance of cause and effect.) In an essay entitled *A Refutation of Deism*, he took on Paley by insisting that "the laws of motion and the properties of matter suffice to account for every phenomenon, or combination of phenomena exhibited in the Universe." This was all humanity *could* understand, he concluded, and the rest was gratuitous speculation.[59]

As she grieved for her firstborn child, Mary Shelley out-radicalized them all. The story of *Frankenstein* and the problem of child loss were in fact tightly intertwined. A few months after her baby died, Shelley recorded in her journal: "Dream that my little baby came to life again; that it had only been cold, and that we rubbed it before the fire, and it lives. I awake and find no baby—I think about the little thing all day." The tale of the vermicelli represented much, namely the idea that life arose via spontaneous generation, purely through the interactions of matter and electricity (what Shelley called galvanism). Shelley imagined that creation and dissolution might be purely mechanical phenomena: the human body, as a machine that arose from natural processes, simply broke down sometimes, and nothing more. Perhaps God had nothing to do with creation, but, more importantly, perhaps He had nothing to do with death.[60]

The Shelleys were ostracized from "respectable" society for their radical beliefs about God, politics, and society. When *Frankenstein* was published in January 1818, plenty of Britons remembered the chaos of the French Revolution and the Reign of Terror. Those in power blamed that event on materialist philosophers who justified dispensing with king and clergymen on the grounds that an "all-good, all-powerful, all-knowing God" did not, in fact, exist. Some, like Erasmus, had talked about a wise, benevolent "first cause" creating species, including human beings, via a purely naturalistic process of evolution, and then used that alternative creation story to insist political and social change must be a "great good" as well. In Britain, such "materialism" was equated with sedition.

Percy lost custody of his first child (by his first marriage) on the grounds, in part, that he was an atheist. Most assumed that atheists could not be trusted to be moral and do their duty since, in repudiating Scripture, they supposedly set aside, first, the Ten Commandments and, second, belief in reward and punishment in the hereafter. Even at the end of the nineteenth century, many remained suspicious of the ability of unbelievers to tell right from wrong, and be moral, upstanding, respectable members of society. For having given up belief in the divine origin of the Ten Commandments and the prospect of heaven and hell, what cause had an atheist to choose to do good rather than evil? Anyone proposing that God either did not exist or was "benevolently" governing from a distance, or that it was impossible for humans to know one way or the other, had to answer this question.

When *Frankenstein* appeared in London bookstalls, Charles Darwin was almost nine years old. Throughout his childhood and well into his adult years, the dominant political, theological, and *scientific* powers within Britain held that both the Book of Nature and the Book of God were on William Paley's side, rather than that of his grandfather or Mary Shelley. It was Paley's *Natural Theology* that was repeatedly published and read during Darwin's youth (with twenty editions by 1820), not Erasmus Darwin's *The Temple of Nature* or Mary Shelley's *Frankenstein*. An 1837 edition by Thomas Smibert made the book available to the "working man." Translations appeared in German, Spanish, French, and Italian, and Paley inspired dozens of other works with titles such as *The Power, Wisdom, and Goodness of God, as Displayed in the Animal Creation*. Nature, in this world, demonstrated the goodness of a closely superintending God, rather than a new vision of nature (and society) improving itself over eons. Meanwhile, Erasmus Darwin's evolutionary vision of creation (and thus his explanation of suffering) was damned as not only unscientific, but atheist, treasonous, antisocial, immoral, and deadly nonsense. There was a lot at stake in choosing between different explanations of the sparrow's fall.[61]

CHAPTER THREE

The Scale of Suffering Expands

CHARLES DARWIN'S FIRST ENCOUNTER WITH PROFOUND HUMAN SUF-
fering occurred shortly after his arrival, in October 1825, in Edinburgh
to study medicine. He witnessed "two very bad operations," one on a
child, and ran away before the operations ended. "Nor did I ever attend
again," he wrote years later, "for hardly any inducement would have been
strong enough to make me do so; this being long before the blessed days
of chloroform. The two cases fairly haunted me for many a long year."
(Chloroform was one of two anesthetics discovered in the early 1840s.)
It was his first experience with how truly awful life on Earth could be (he
had lost his mother to a terrible illness as a boy, but was not allowed in
her sick room until it was over).[1]

When Darwin began medical school, most of his professors defended
the study of human anatomy as useful in the defense of Anglican interpretations of God's governance. Professor of anatomy Alexander Monro,
for example, taught anatomy "to get a view of the animal creation, which
affords such striking illustration of the wisdom and power of its Author."
There were some, however, who were drawing very different conclusions
from the study of the human body. Some anatomists were even calling
upon evolutionary explanations of species as a means of breaking the
power of the Anglican Church. As we will see, at least one of these men,
Dr. Robert Grant, expected the grandson of Erasmus Darwin to sympathize with such views. But Charles Darwin did not find his grandfather's
radical ideas about either God or nature appealing during his time in
Edinburgh. He initially found no reason to adopt Erasmus's fantastical,

perhaps even seditious, evolutionary account of the origin of species nor his law-bound explanation of the fall of sparrows.[2]

Indeed, when he left medical school (without taking a degree) in 1827, Darwin enrolled in Cambridge University to train for the Anglican ministry. In doing so, he not only chose Paley's explanation of the origin of species, he took sides in a debate over why suffering exists in the world. Why wasn't Darwin initially interested in evolutionary explanations of either the origin of species or the existence of suffering? Answering that question requires that we examine where Darwin fit within a spectrum of political, medical, scientific, and theological positions that existed by the 1820s. Only then will we be able to track why, eventually, he changed his mind about so much.

I Listened with Silent Astonishment

Born in 1809 to wealth, privilege, and respectability, Charles Darwin had four career options: medicine, the law, the ministry, or the military. His father thought he would make a good physician so Charles went off to Edinburgh at the age of sixteen, in the company of his older brother, Erasmus Alvey Darwin.

Medical training in Edinburgh was in a bit of an uproar when Darwin arrived. Wealthy, established physicians were struggling for control over the future of the profession with upstart reformers out to change the world, and change it quickly. Two kinds of medical schools vied for students' allegiance. The comfortably established, upper-class medical professors at the university called upon Paley's version of the design argument to seal the case for God's close personal providence over society. The other kind of medical school existed outside the control of the university. These "extramural" schools tended to attract religious dissenters who were uninterested in justifying the existing political or social structures.

Given the use of the theory of special creation to demonstrate God's close personal control over the hierarchical structure of British society (including the medical profession), alternative explanations of the origin of species proved appealing to individuals who desired political, economic, and social reform. Indeed, historian Adrian Desmond notes that a commitment to radical reform was the common denominator among physicians who adopted theories of evolution. Radicals and reformers found evolution appealing for two reasons: First, distancing God's hand from both nature and society undermined arguments that the political, social, and economic status quo originated from God's design. Second, a vision of nature that made large-scale change *natural* seemed to justify arguments that society could and should be changed.[3]

Medical men on the margins of political power had both the incentives and the anatomical knowledge to challenge establishment anatomists like Professor Monro and their grand statements about the wisdom and power of the Creator. They drew, for example, on the study of teratology (or "monstrous" births) to argue that abnormal organs were the result, not of God's close personal providence, but of undesigned errors in the law-bound process by which matter developed from conception to an adult organism. And once they made that move, driving natural law into the distant past, the origin of species, and even the origin of life, easily followed.[4]

Charles Darwin learned the connection Edinburgh medical men were drawing among politics, anatomical knowledge, and the idea of evolution from some of the most radical anatomists in Edinburgh. One of his mentors, Dr. Robert Grant, had spent summers in Paris learning directly from transmutationist and comparative anatomist Étienne Geoffroy Saint-Hilaire. Geoffroy described vertebrate species as variations (via a purely natural process of divergence) of one unified, original plan designed by a distant first cause. Indeed, he defended the radical, materialist work of his colleague at the National Museum of Natural History in Paris, Jean-Baptiste Lamarck, who had developed a materialistic explanation of the origin of purposeful parts (something William Paley had argued was impossible). Like Erasmus Darwin, Lamarck assumed that modifications arising from increased use or disuse during an individual's lifetime could be inherited by the young. If useful, those modifications would persist and increase over time. Here, for example, is Lamarck's account of the origin of the webbed feet of waterfowl:

The bird attracted by need to the water to find there the prey necessary for its existence, spreads the digits of its feet when it wishes to strike the water and move on the surface. The skin that unites these digits at their base thereby acquires the habit of stretching itself. Thus, with time, the large membranes uniting the digits of ducks, geese, etc. have been formed such as we see them today.[5]

Lamarck chose the example of birds' feet not because he was an expert in ornithology (he was actually an invertebrate zoologist), but because other authors, including Paley, cited the diversity of bird feet as demonstrations of God's special, independent creation. Lamarck, by contrast, insisted that purposeful parts like webbed feet could be explained using a purely materialistic, mechanical process. Like Erasmus, he also imagined life could arise via spontaneous generation. He saw no need to posit a miraculous (i.e., supernatural, immaterialist) origin of species: nature could do all.

Robert Grant agreed with both Geoffroy and Lamarck's ideas about the origin of species. As we have seen, there was a strong political, as well as theological, edge to that agreement. Grant would eventually be ostracized from respectable society for his political radicalism. He loved Lamarck's vision of an upward, striving, changing nature, in part because it seemed to put society in motion as well. After all, Lamarck imagined organisms themselves producing wondrous progressive change "from below" rather than social hierarchies arising from "above" (i.e., from God). Under the assumption that what is natural is good, the lesson for those oppressed by hierarchy and authoritarian monarchies and clergy (or medical boards) seemed clear: activist campaigns to change the world were approved by nature and (albeit a distant) God. Conservatives countered that the presence of hierarchy and design in nature demonstrated the exact opposite.

When Darwin learned of Grant's belief in transmutation, then, he was also learning about Grant's politics. "He, one day," Darwin later recalled, "when we were walking together burst forth in high admiration of Lamarck and his views on evolution. I listened with silent astonishment, and as far as I can judge, without any effect on my mind."[6]

It turns out that Grant had first become enamored of evolutionary visions of nature when reading Erasmus Darwin's work on the "laws of organic life." Historian Janet Browne notes that he was probably delighted by the fact his young friend was a "direct descendant of this famous English evolutionist. Perhaps he expected Darwin to be equally unorthodox, predisposed to discuss transmutation with the same airy directness of his grandfather." If so, he was disappointed. By this time Charles had read the text of a lecture on transmutation by Lamarck. He had also read and "admired greatly" his grandfather's book *Zoonomia*. But he saw no reason to adopt the radical evolutionary ideas that peppered his grandfather's poems or medical books.[7]

Indeed, Darwin had plenty of political, scientific, and theological reasons *not* to have anything to do with such ideas. As we have seen, the men in his family (on both the paternal Darwin side and maternal Wedgwood side) had a strong tendency toward radical political and theological thought. Yet by the opening decades of the nineteenth century, the Darwin-Wedgwood clan had settled into a comfortable respectability rooted in liberal (rather than radical) attitudes toward politics, economics, and religion. (The term *liberal* in the nineteenth century generally meant support for social and economic reform in the interest of middle-class industrial capitalists, a free market economy with limited government interference, and relative freedom of religion.) In the wake of the French Revolution, British politics had solidified into two major parties: the conservative Tories committed to the institutions of the past, including both the monarchy and the Church of England, and the liberal (but by no means radical) Whigs, who argued for targeted reforms. The Darwins (and Wedgwoods) were upper-class, respectable Whigs, committed to social, economic, and political reform. Whigs were willing to change the status quo (up to a point) for two reasons: first, to prevent calls for revolution from crossing the channel; and second, to increase the political and social power of their own rising industrial-capitalist class.

By the time Charles was born, his family had largely set aside Erasmus's radicalism. As wealthy beneficiaries of the Industrial Revolution and capitalism (Darwin's uncle, Josiah Wedgwood, was one of the wealthiest industrialists in Britain), the Darwin-Wedgwood clan had very little to gain from political and economic revolution. And by this point in time the idea of evolution had very clearly been harnessed to more radical calls for social and political change—in Paris, Edinburgh, and London—than the Darwin family could countenance.

There was one issue on which the Darwin-Wedgwood clan took a more radical stance (in calling for immediate, rather than slow, change). They were passionate abolitionists who campaigned for the immediate, rather than gradual, abolition of slavery. Would not the idea of evolution appeal to those calling for slavery to end and keen to demonstrate that all men are brethren? Clearly grandfather Erasmus had seen an evolutionary path toward a resounding "YES!" in reply to (as he imagined) the enslaved man's query, "Am I not a Man and a Brother?" But Erasmus was an outlier; few abolitionists adopted his evolutionary defense of human unity. Indeed, as we will see, for much of the nineteenth century it was not clear whether the idea of evolution would serve the abolitionist (or any other humanitarian) cause. Two decades after Erasmus's egalitarian poetry, most abolitionists still centered their arguments about the "unity of man" in Scripture. Eventually, Darwin's abolitionist background would play a crucial role in his conversion to evolution, but in the 1820s he did not consider his grandfather's speculative defense of human unity to be an appealing replacement for the Bible's claim that "God hath made of one blood all the nations of men" (Acts 17:26).

And what about religion? Theologically, the Darwin-Wedgwood clan had adopted a complicated mix of more conservative Anglican and more liberal Unitarian practices and creeds. This meant that, by the time Charles was in Edinburgh, he had access to a host of alternatives for how to think about God's governance, including the origin and meaning of suffering. The Anglican side of the family held that pain, suffering, and death originated as just punishment for Adam and Eve's disobedience. They held that the wicked would be punished for their sins, and that Christ's atonement for human sinfulness on the Cross offered hope to the faithful of salvation in heaven. The Unitarian side repudiated belief in hell while trying to maintain belief in heaven, and set aside orthodox interpretations of Genesis (especially original sin and the fall of man) as explanations of the human condition.

As Unitarians set aside Anglican explanations of the past, they cast about for alternative histories and explanations of human nature. Again, it might seem like evolution would fit the bill (certainly grandfather Erasmus thought so), but by the 1820s it was not clear that evolution was a defensible *rational* or *scientific* proposition. And Unitarians wanted their explanations to be both rational and scientific. The most renowned naturalist of the era, the French comparative anatomist Georges Cuvier, had ridiculed evolutionary explanations as beyond the boundaries of good science. He famously dismissed Lamarck's evolutionary ideas with disdain in an elegy after Lamarck's death. The theory that some parts might change through use and disuse and create new species, Cuvier warned, "may amuse the imagination of a poet; a metaphysician may derive from it an entirely new series of systems; but it cannot for a moment bear the examination of anyone who has dissected a hand, a viscus, or even a feather." Cuvier argued that the careful study of anatomy proved the impossibility of one species changing into another species. As a strict empiricist (someone who demanded observational evidence for claims), he insisted that scientific naturalists must simply take species as created, somehow, in the distant past, and focus on what could be observed in the present. And what naturalists observed was extraordinary functional correlation, limited variation, and perfect adaptation to particular environments.[8]

Darwin may not have learned much medicine in Edinburgh (he would later regret his lack of training in dissection as an "irremediable evil for all my future work"). But while avoiding the operation theatre, he made friends, attended natural history society meetings, and went on field trips to collect natural history specimens. As he did so, he learned a great deal regarding the political, scientific, and theological status of evolutionary visions of the origin of species. He also learned how disturbing radical, materialist ideas about the origin of life (such as spontaneous generation) could be to devout individuals trying to make sense of the world. His good friend John Coldstream became plagued by religious doubts soon after graduating, doubts that his memoirist attributed to "certain Materialist views, which are, alas! too common among medical students." Coldstream himself attributed his religious doubts to the doctrine that disbelievers would be condemned by God to eternal misery, but the fact his nervous breakdown was attributed to "materialist views" indicates the danger perceived in ideas circulating among some of the medical men.[9]

Coldstream eventually recovered his faith in both a "merciful and gracious" God who "never visits his children with Affliction in vain" and "the good hope of a blessed immortality." As a physician, he also continued to argue that the study of both natural history and anatomy supported Christian faith. During illness, the chirping of the sparrows reminded him of "the providential care of his heavenly Father." One might meditate on the lines about sparrows in Matthew. Or one might ponder the extraordinary design of the sparrow's beak, so wonderfully designed for procuring sustenance. In either case, Coldstream believed the sparrows taught a united message: God's close, personal control over human lives. He once expressed what he feared might be lost, if one abandoned such belief, when returning home from the funeral of a child: "Could we not look upwards and say 'the Lord reigneth,' . . . such sorrows would be overwhelming." By all accounts, when he left Edinburgh, Darwin agreed with Coldstream, as did the majority of British naturalists.[10]

In this world, the rational study of natural history, as the entomologist William Kirby wrote, demonstrated that "nothing really happens by chance, or is the result of accidental fortuitous events: second causes are always under the direction of the *first*, who ordereth all things according to the good pleasure of his will." Geologists might describe the elevation of a new island or a volcano by natural laws, but phenomena "still may be denominated [by] *his* work . . . When we are assured that the hairs of our head are all numbered, and that not a sparrow falleth without our Heavenly Father, we are instructed to look beyond second causes for the direction and management of events." The Oxford geologist Reverend William Buckland even traced design, benevolent contrivance, and the "finger of an Omnipotent Architect providing for the daily wants of its rational inhabitants" in the minerals, metals, salts, ores, and coal hidden deep within the earth, and "accessible at a certain expense of human skill and industry."[11]

In an age of rising industrial wealth, stark economic inequality, and unregulated child labor, Reverend Buckland's argument would, as we will see, prove dangerous. For as the historian and philosopher of science

Michael Ruse notes, obviously the authors of such statements had never "worked in a pit, nor did their children have to." For those less inclined to see God's personal hand in human suffering, such arguments seemed an all-too-convenient justification of an unjust status quo. They were also a profound reminder that if one wanted to change the world, natural history's tight ties to particular theological and political claims might need to be changed too.[12]

THE NATURE AND LAWS OF THE INFANT CONSTITUTION

Darwin spent much of his time in Edinburgh dissecting sea creatures and learning how to stuff birds, but he was surrounded by reformers and radicals asking why the world is the way it is and whether it could be changed. What, they asked, is the proper scope of human action in a world of divinely designed natural laws? What must be done, if anything, when nature's laws seem to result in suffering? These questions haunted a young Scottish physician named Andrew Combe, who had just commenced practicing medicine in the city. Combe watched debates over natural law, the divine will, and human action closely, especially fights inspired by the work of an Anglican reverend named Thomas Malthus. It is worth spending time with both Combe and Malthus, for their explanations of child mortality would eventually have a tremendous influence on Darwin's vision of both nature and God.

Reverend Thomas Malthus's 1798 book *An Essay on the Principle of Population, as It Affects the Future Development of Society* is known today primarily for the pivotal role it later played in Darwin's discovery of natural selection. (Upon reading the sixth edition of Malthus's book in 1838, Darwin finally hit upon his own explanation for the origin of purposeful parts.) But for this story, Malthus is important because he set the terms for what British debates over the origin of suffering, including child mortality, looked like by the time Darwin was at medical school. Malthus wrote *An Essay on Population*—first, to explain why God made the world the way it is, and second, to take a stand on whether, how, and to what extent the earth might be changed by human effort. Malthus knew from long experience with his parishioners that a "fearful mortality" existed among infants. And he suspected that many of these deaths (including the desperate act of infanticide) resulted from the fact that a portion of the people were underfed.[13]

Why did poverty, which caused so much suffering and perhaps even higher rates of child loss, exist if God was so bountiful and good? Malthus knew that some blamed the class system. They viewed that system as an artificial, unjust creation of the wealthy, and in the wake of the French Revolution, they proposed radical solutions aimed at leveling society and ameliorating the suffering of the poor and underfed. But Malthus was certain that such schemes relied far too much on visions of what the world *should* be like, with very little attention to human nature and nature's laws. He was particularly worried by the ideas of William Godwin, Mary Shelley's father, who argued that an egalitarian society could be produced in which *all* individuals prospered.

Godwin had imagined that society could be reorganized in such a way that all would have enough to eat. Malthus was certain that such utopian schemes ignored two divinely instituted natural laws: first, food is necessary to human existence; second, "the passion between the sexes" is both necessary and unchangeable. Because, by Malthus's calculation, the population increases about four times faster than agricultural production, an inevitable struggle for existence *always, inevitably, and necessarily* follows prosperity. Humans, Malthus concluded, will always "breed up" to a point where resources are limited. Thus, a struggle for existence will always arise, and poverty and suffering result. Increase charity for the poor, he argued, and the population would just increase faster since the poor would have no incentives (via the threat of starvation) to delay marriage.

Malthus then described two kinds of "checks" on human population growth, given the above "laws": preventive checks (i.e., those that *prevented* births, including the "vices" of prostitution and contraception) and positive, or "active," checks (those that *increased* premature death, including disease, starvation, and war). Given these inevitable checks (all of which, he argued, cause misery), Malthus argued that misguided charity would, by increasing the population, *increase* the amount of suffering in the long term. "True charity," for Malthus, must be based upon the premise that no child should be eligible for parish relief. Critics pointed out that such a policy was surely "hard on the poor children, who came into the world by no fault of their own," but Malthus thought the existing system (which gave families relief based on the price of bread and number of children) far worse, if only one looked closely enough.[14]

Though he would be pilloried by his critics as hard-hearted, Malthus was driven by his firm belief that the poor already suffered enough. Since 1788 he had served as curate (or minister) of the village of Okewood, where many of his parishioners were poor laborers. He knew that their children went about without shoes or stockings, that two-fifths of their income was spent on bread alone, and that they lived in leaky cottages with few comforts and much smoke. He knew that babies came again and again because he baptized the little boys and girls and entered them into the Okewood Register. He also recorded the deaths (when he knew about them). One couple he married would have thirteen children in all, four of whom died under the age of six months.[15]

Having described the "laws of population," Malthus apologized for the "melancholy hue" of his book. But he insisted that he drew "these dark tints, from a conviction that they are really in the picture, and not from a jaundiced eye or an inherent spleen of disposition." In response to critics who held he had impiously overdrawn the severity and constancy of the struggle for existence, he added (to subsequent editions of his *Essay*) a 150-page catalogue of "poor food, lack of ability to care for the young, scant resources, famine, infanticide, war, massacre, plunder, slavery, cold, hunger, disease, epidemics, plague and abortion" in human societies.[16]

Malthus cited natural laws to explain poverty, famine, and thus, implicitly, differential child mortality. But in what sense, critics asked, did a providential purpose and meaning exist in each individual loss if all happened due to natural laws? What had become (in the words of one of Malthus's critics, Dr. Jarrold) of "a wise and benevolent Creator (who) has his eyes constantly upon us?" Another insisted it was "impious and atheistical" to assert that "the Almighty brings more beings into the world than he prepares nourishment for," and demanded to know whether there was no law "in this kingdom for punishing a man for publishing a libel against the Almighty himself?"[17]

William Paley, whom we met in chapter 1, agreed that explaining the existence of evil via natural laws, as Malthus did, was difficult to apply "in individual cases." But, in contrast to Malthus's critics, Paley enthusiastically adopted Malthus's explanation of poverty as resulting from benevolent, divinely designed natural laws (Malthus's essay appeared four years before *Natural Theology*, in 1798). Searching for purpose in every affliction, Paley believed Malthus was explaining human experience rather than impugning God's goodness. Indeed, both Paley and Malthus thought that, in describing the "general laws" through which God chose to govern, they were *recovering* God's benevolence, wisdom, and power despite the great suffering caused by poverty.[18]

One of the most complicated (and, as we will see, often forgotten) components of both Paley and Malthus's use of these "general laws" is that they saw themselves as explaining the means God had provided to ensure human moral and material *progress*. Malthus, for example, wrote of natural laws as arranged by God, who "still executes, according to fixed laws, all its various operations" for "the advantage of his creatures." The evils that inevitably resulted from these "fixed laws" were, Malthus argued, in fact benevolent "instruments employed by the Deity in admonishing us to avoid any mode of conduct which is not suited to our being, and will consequently injure our happiness." For example, without the threat of famine due to the population and food increasing at such different rates, man would have "no motive . . . sufficiently strong" to overcome his "indolence" and work to cultivate the soil. Viewed from the right perspective, then, Malthus insisted that the "laws of nature, which are the laws of God," produced more good than evil.[19]

As we saw in chapter 1, a long tradition existed in British science of seeing ignorance of divinely designed natural laws as the source of much misery and suffering. Like the first codifier of scientific methods, Francis Bacon, Malthus held that the fact humans might understand these laws and thus ameliorate suffering in future constituted a benevolent provision of God. And like both Paley and the seventeenth-century physician Thomas Sydenham, Malthus argued that the constancy of natural laws provided "the foundation of the industry and foresight of the husbandman, the indefatigable ingenuity of the artificer, the skillful researches of the physician and anatomist, and the watchful observation and patient investigation of the natural philosopher. To this constancy we owe all the greatest and noblest efforts of intellect. To this constancy we owe the immortal mind of a Newton." Had humanity been able to rely on miracles, Malthus concluded, they would have become slothful and complacent. And *that* was why God, in his infinite wisdom, did not intervene (say, by answering prayer with a miracle), even amid the worst afflictions.[20]

By all accounts an earnest Anglican, Thomas Malthus believed in the God who attended the fall of every sparrow. Like many Anglicans, he reconciled his emphasis on God's governance via natural laws with his belief in a personal God by emphasizing a distinction between general providence (divinely designed natural laws) and special providence (God's use of those laws for particular purposes). In other words, God, who was all-powerful and all-knowing, governed primarily via natural laws. Thus, Anglican ministers could write about natural laws, yet still urge those afflicted with disease or loss to look beyond secondary causes to the ultimate, ever-present source of events. Meanwhile, given the strong consensus regarding the miraculous, independent creation of species, the afflicted could ponder the extraordinary design of the hinges of an earwig's wing, and thus be reminded of God's attention to even the smallest of creatures.

By the 1820s Paley and Malthus's talk of a beneficent, divinely designed struggle for existence that arose from the "fixed laws of our nature" had permeated British (and American) natural history and political economy. Yet it soon became clear that a profound tension lurked within their explanation of the world and God's ways: if progress depended upon a divinely designed struggle for existence, then perhaps trying to prevent one of its most obvious consequences, high child mortality, might constitute misguided interference in God's plan. This was *not* Malthus's point: he believed suffering could and should be ameliorated via human effort, but that to do so wisely man must use his God-given reason to understand nature's ways (rather than ignore both human nature and natural laws, as he believed the French had done). Indeed, Malthus explained suffering as the necessary driver of a divinely approved yet arduous path to amelioration and progress. "We shall never be able to throw down the ladder," Malthus wrote, "by which we have risen to this eminence; but it by no means proves, that we may not rise higher by the same means." Both Paley and Malthus argued that the fact "population always treads on the heels of improvement" should not dishearten endeavors for "public service."[21]

It did not take long, however, for some to find in Malthus's "laws of population" a convenient excuse for inaction in the face of great suffering. Indeed, having turned the "struggle for existence" into a struggle between "the races," defenders of slavery in the United States tried to put God and nature on their side by citing Malthusian laws. They claimed, for example, that slavery *protected* the descendants of Africa from "inevitable extermination." Such arguments relied upon the pervasive assumption among those of European descent that whites would always out-compete anyone else amid the "struggle for existence." Some even used Malthus's work to argue that slave economies produced *less* suffering than free labor systems by making "laborers" subject to their "masters," rather than their "passions."[22]

Even those who thought slavery was a moral evil cited Malthus as an excuse to ignore the suffering of men, women, and children of African descent. The theologian Horace Bushnell, for example, cited Malthus's work as evidence that abolition (which he supported) would, in giving free reign to the inevitable "struggle for existence," lead to the extinction of the African "race." Bushnell did not believe anything could or should be done to prevent either the struggle or extinction. Indeed, he wrote that he saw the hand of "the Almighty Himself" at work in nature's ways: "Since we must all die," Bushnell wrote, "why should it grieve us, that a stock thousands of years behind, in the scale of culture, should die with fewer and still fewer children to succeed, till finally the whole succession remains in the more cultivated race?" Others, as we will see, removed "Almighty God" from the equation entirely while keeping the conclusion that both the struggle and extinction were great goods.[23]

Thomas Malthus was dismayed by how the pro-slavery faction used his *Essay*. As an abolitionist who believed in the unity of man and in both material and moral progress as divinely approved, Malthus had not meant to halt attempts to better the world. But clearly by the 1820s, when Andrew Combe was practicing medicine, Malthus's *Essay on Population* was being used to argue against any attempt to improve social and economic conditions for the poor. And so, while Darwin avoided the surgical operating theatres, Combe fought back against those who cited Malthus as evidence that high mortality among the young "constitutes a necessary part of the arrangements of Divine Providence which man can do nothing to modify."[24]

Having specialized in the medical care of children, Combe was adamant that parents and physicians must see the current state of affairs as something that required *action* rather than resignation. High mortality, he argued, proceeded "chiefly from secondary causes purposely left, to a considerable extent, under our own control." If this was true, then the key to saving children from death and disease was to teach parents that their own inattention to natural laws (and not God, whether working through secondary causes or not) doomed beloved children to an early death. Only when parents saw illness as the result of *purely* natural cause and effect, he insisted, would they make the changes required to keep infants alive.[25]

On this point, Malthus would surely have agreed with him. But Combe imagined far more radical reforms than Malthus was willing to countenance. Like Malthus, Combe knew that between a third and a half of all children died before the age of five. But Combe was most interested in the fact that, upon closer examination, that rate *varied* according to conditions: the children of the poor died at higher rates than those of the rich; those of densely peopled manufacturing towns at higher rates than those in the open country; and those in foundling hospitals (orphanages) at enormously higher rates than those within families. "According to the existing constitution of society," Combe wrote, "many of the comforts, and some even of the necessaries of life, are beyond the reach of the poorer and working classes; and this circumstance will be found to operate unfavourably in diminishing the chances of infant life amongst them." The operative phrase in Combe's argument here is "according to the existing constitution of society." For Combe believed the "constitution of society" could and should be changed, and changed radically.[26]

Combe's biographers could find no evidence he believed in Christianity, but his language shows how the language of natural theology could be harnessed by both Christians (who believed in a close personal God) and deists (who moved God to a distance and did not believe in the New Testament). In urging parents to amass the requisite knowledge to save children, for example, Combe described how God, in his infinite omniscience, creative power, and beneficent wisdom, had arranged the structure and laws of the infant body "for our welfare and advantage," and that therefore parents must study these laws to avoid bringing disease and suffering upon their children. In proportion as "we shall discover and fulfill the laws which the Creator has established for our guidance and preservation," so, Combe argued, would the terrible waste of infant life decline. He wrote:

> *When the principles of treatment are traced back to their true and only solid foundation in the nature and laws of the infant constitution, the rules of conduct deducible from them come before us stamped with more than human authority. In so far as they are accurately observed and truly recorded, they assume the dignity of positive, though indirect, intimations of the Divine will, and claim from us the same reverential obedience as all the other acknowledged commands of the Almighty.* [27]

Like Malthus and Paley, Combe premised his system on the absolute uniformity of natural laws. He urged his readers to see the great unchangeable laws of the Creator as a benevolent dispensation. Only if natural laws were uniform and unchangeable could cause and effect be discerned: to believe otherwise was to shut our eyes to the means by which God governs, and then presumptuously pray that He would alter the order of nature in our favor and "remove the consequence of our deliberate disobedience." But in contrast to Paley and Malthus, Combe made no effort to maintain God's close personal providence. He did not urge grieving parents to look up to the first cause when medicine failed. And he did not direct them to the extraordinary design of swallows or woodpeckers to remind them that *"one of them shall not fall on the ground without your Father."*[28]

We have seen how Robert Grant drove secondary causes into the deep past by positing an evolutionary explanation for the origin of species. To imagine purposeful parts arose via some purely naturalistic process, rather than through independent creation, was useful to radicals like Grant, who wished for society to be completely remade. Andrew Combe was not interested in supporting Anglican ministers, physicians, naturalists, and anatomists' use of the purposeful parts of animals and plants to uphold the "existing constitution of society" as providential. He was also well aware of the fact that transmutationists were removing the "theory of independent creation" as a prop for existing social structures. He never wrote about transmutation, however, except to say that as a busy physician trying to save lives in the present, he did not have time to speculate about the origin of species. But clearly someone else, with a little more time, might.[29]

THE STUDY OF GOD'S WORKS

In the 1820s Charles Darwin was not that "someone else," and there were no indications that he might be in future. He was perfectly satisfied with Paley's complicated negotiation among secondary causes, the theory of independent creation, and belief in God's personal providence. During his time in Edinburgh, Darwin's religious, theological, and scientific sympathies were entirely with the critics of Lamarck and Robert Grant. There is no evidence he had seen any reason, witnessed any evidence, or heard any arguments that had changed his mind, about either species or God, when he left Edinburgh in 1827.

Meanwhile, Darwin's decision to flee the operating theatre need not have ended his medical career. A country doctor did not have to bloody his hands with surgery (which was always a last resort given the high mortality rates from infection and lack of anesthetics). As his father told him, the chief element of success with patients was the ability to inspire confidence. Most patients just needed to talk. But Charles had had enough of medicine. Besides, as he recalled years later, his discovery that he would inherit plenty to live on while maintaining his status as a "gentleman, dampened "any strenuous effort to learn medicine."[30]

As a result, Darwin returned home without a medical degree and confronted a very disappointed father. Robert Darwin had already exclaimed, "You care for nothing but shooting, dogs, and rat-catching and you will be a disgrace to yourself and all your family." By all accounts a dutiful son, young Darwin was mortified. He knew he had to choose between one of the other three careers appropriate for a gentleman. Uninterested in the military or law, Darwin left for Cambridge to study for the Anglican ministry. For a young man who loved natural history, it was definitely the best career option after he decided against medicine. Thanks to John Ray, William Paley, and generations of naturalists, Darwin could imagine collecting beetles on weekdays and composing sermons for Sunday (even if his elder brother Erasmus, who had quite radical tendencies, chided him for his choice).[31]

Clearly the radical critics of the theory of special creation had thus far had little impact on Darwin's thinking, except to convince him that evolution was not a respectable scientific, political, or theological option for a man of his standing or interests. Medical reformers like Robert Grant might ridicule Paleyan anatomists' pious descriptions of God's benevolent design. They might rebel against the ease with which natural theology could be used to justify a status quo so full of misery and suffering for some. But Darwin had little reason to rebel against the privileged class into which he was born (so the political utility of evolution did not appeal), and evolutionists had little to offer in the way of a convincing mechanism through which either species or purposeful parts might form (so on scientific grounds, evolutionary explanations of species failed).

The sincerity of Darwin's commitment to becoming an Anglican minister has been doubted by those who wish to see the seeds of rebellion as early as possible. But there is no evidence in the historical record that his interest in the ministry was insincere. He did ask for time to consider, for, as he later recalled, "from what little I had heard and thought on the subject I had scruples about declaring my belief in all the dogmas of the Church of England; though otherwise I liked the thought of being a country clergyman." (As of 1828 the Test and Corporation Acts allowed

dissenters to hold public office, but they still could not perform the sacrament of marriage or burials, and only Anglicans could attend Oxford
and Cambridge.) So, he read "with care" *Pearson on the Creed* "and a few
other books on divinity," and persuaded himself "our Creed" (that is, the
creed of the Anglican church) must be fully accepted.[32]

Pearson on the Creed was a two-volume book composed in the seventeenth century by the Lord Bishop of Chester, John Pearson (it was
originally entitled *An Exposition of the Creed*). Pearson explained every
one of the Thirty-Nine Articles of the Anglican Church in great detail,
so that everyone, "when he pronounceth the Creed, may know what he
ought to intend, and what he is understood to profess." In his discussion of Article I, for example (God is "the Father Almighty, Maker of
Heaven and Earth"), Pearson cited the authority of both "the wisdom
of the Jews and St. Paul" for the claim that a knowledge of Creation
(i.e., natural history) led human beings to "a clear acknowledgement of
the supreme and independent Being." In nature, for example, one could
"read the great Artificer of the World in the works of his own hands, and
by the existence of any thing demonstrate the first cause of all things."
The point of this claim became explicit in the following line: "God had
as absolute power and dominance over every person, over every Nation
and Kingdom on the earth, as the potter hath over the pot he maketh,
or the clay he mouldeth. Thus are we wholly at the disposal of his will,
and our present and future condition framed and ordered by his free, but
wise and just, decrees." This faith in God, "whose will is a law to us, cannot do anything unwisely or unjustly," and, as Pearson explained, it was
necessary both to direct our actions and "for consolation in the worst of
conditions."[33]

Darwin's belief, for now, that this creed must be fully accepted
explains why, as he put it, he was charmed by Paley's argument when he
read *Natural Theology* as a student at Cambridge. "I did not at the time,"
Darwin added, "trouble myself about Paley's premises." Those premises
included the belief that the close personal God of Christianity exists; that
God is all-wise, all-benevolent, and all-powerful; and that both Scripture
and nature were adequate routes toward demonstrating the truths of
Christianity.[34]

By this time Darwin knew the alternatives to the theory of special creation well enough—after all, he had spent two years in Edinburgh listening to debates about materialist accounts of life, read his grandfather's works, and walked endless hours with transmutationist Robert Grant. But in enrolling in Cambridge he cast his lot with the theory of special creation. (As we will see, the social and political allegiances of that choice never left him, even after, much later, he set aside the God who attended to the fall of every sparrow.) During his two years in Edinburgh, Darwin had learned that, even as the realm of natural law expanded, most professional naturalists and physicians viewed transmutation as extremely speculative (and thus nonscientific), and both politically and theologically dangerous. As firm proponents of the theory of special creation, Charles's favorite professors at Cambridge, the botanist Reverend John Henslow and the geologist Reverend Adam Sedgwick, reinforced this consensus. Both men, whom Darwin respected tremendously, taught that the study of God's "two books"—Scripture and Nature—went hand in hand.

John Henslow's biographer attested to the fact his "ministerial teaching was in strict agreement with the Sacred Volume, as it was also with the Articles of our Church."[35] Darwin recalled that Henslow was "so orthodox that he told me one day, he should be grieved if a single word of the Thirty-Nine Articles were altered."[36] The same month that he and his wife Harriet lost their six-month-old son John Jenyns, Henslow delivered a sermon in which he spoke of his belief in the strictly literal truth of the two resurrections prophesied in the Book of Revelation. At that great event, Henslow preached, "the dead both small and great are to stand before God." He firmly believed that those who became the willing slaves of Satan in this world, by abandoning themselves to vicious practices and unbelief, would be citizens of hell in the hereafter.[37]

Well aware that Unitarians insisted a good God could not damn even the wicked for eternity, Henslow insisted that "carnal understanding" (i.e., human reasoning, "clogged with the flesh") could not understand such things. The prospect of eternal damnation, he argued, represented a spiritual truth to which humans must respond with prayer, "and then God will in his own time reveal so much of his mysteries as he may himself think fit." Clearly there was much, to Henslow, that humans could not understand according to their own standards of justice. Henslow did believe that as a moral being endowed by God with a conscience, one could state one's obligations on Earth with confidence: for Henslow, to be a Christian was to do good and work for justice. His biographer recorded how "in seasons of particular distress and hardship, he would enforce the moral right which the sufferers had to relief from their richer neighbors; at the same time reminding his hearers of an over-ruling Providence, whose care extended to rich and poor alike, and who ordered all events for the good of those who look up to and trust in God."[38]

Radicals ridiculed Anglican Christianity for what they called quietism (i.e., acceptance of things as they are, on the grounds that events are all due to God's will), but Henslow clearly believed reform could coexist with orthodoxy. He was an abolitionist. He fought with the landowners in his parish to allot plots to tenants and educate the children of the poor. He was, Darwin wrote, roused by a "bad action to the warmest indignation and prompt action." Once, when they were walking together in Cambridge, they came upon a "crowd of the roughest men" dragging two body-snatchers (men who dug up fresh graves to supply medical schools with bodies for dissection) by their legs along the muddy and stony road. "They were covered from head to foot with mud and their faces were bleeding either from having been kicked or from the stones," Darwin recalled. "They looked like corpses, but the crowd was so dense that I got only a few momentary glimpses of the wretched creatures. Never in my life have I seen such wrath painted on a man's face, as was shown by Henslow at this horrid scene. He tried repeatedly to penetrate the mob; but it was simply impossible. He then rushed away to the mayor, telling me not to follow him, to get more policemen."[39]

Henslow knew what he was demanding his parishioners believe when urging them to find evidence of the wisdom, power, and goodness of God in the extraordinary design of a flower. In contrast to Professor Sedgwick, who never married, Henslow married and experienced the loss of a beloved child at least once. He and Harriet had a young family when Darwin arrived in Cambridge (Harriet eventually bore six children). The son who died at the age of six months died the same year Darwin first attended Henslow's lectures and soirees (in March 1829).

Henslow's thoughts on his son's death are, as often happens, lost to the historical record (as are Harriet's). But we can guess, based upon his role as an Anglican minister, how he believed one must respond to such loss: faith that, in the words of the first article of the Thirty-Nine Articles of Religion, God is of "infinite power, wisdom, and goodness; the Maker, and Preserver of all things both visible and invisible."[40]

Henslow was remembered by his parishioners as someone who would visit a dying child, offering comfort if he could, every day until they were gone. He firmly believed that, even amid such great affliction, natural history was a crucial means, with Scripture, of revealing God's attributes insofar as fallible human reason could do so. Later in life, as parish priest of Hitcham, Henslow developed rigorous botanical classes for the children of his parish. "Objects of Natural History, Animal, Vegetable, and Mineral Kingdoms" would be exhibited, explained the outline of the lessons, "and accounts given of them as may tend to improve our means of better appreciating the wisdom, power, and goodness of the Creator."[41]

Some complained that he spent too much time on botany and too little on Scripture, but Henslow saw the knowledge obtained from both Scripture and nature as tightly intertwined. Indeed, he blamed the view that science must lead to unbelief on ignorant ministers. Properly understood, it was impossible, he insisted, "that either the works of God, or the Word of God, can ever be teaching us things contradictory to truth." He preached that a knowledge of God's works helped man understand God's providence, and improved the light by which the interpretation of Scripture could be corrected. Henslow was certain that "these two witnesses to His infinite power, wisdom, and goodness, may mutually corroborate each other," and, most importantly, that the study of God's works could not "deprive us of the enjoyment of one jot or tittle of those glorious promises which have assured to us a blessed immortality." To that end he defended and campaigned for more support for the teaching of natural sciences in universities.[42]

Henslow worked within a long-standing British tradition of science that held that the only means of producing knowledge about nature, given man's prejudices, pride, and biased desires, was strict empiricism. So long as observation was disciplined and careful, the study of nature could not, he believed, be in conflict with Scripture. For Henslow, increasing true knowledge about nature was a form of revelation.

Darwin was thus trained by naturalists who knew that discoveries about the natural world might require orthodox interpretations of the Bible to be changed. After all, Anglicans had long since adopted Copernicus and Galileo's new model of the cosmos and made Isaac Newton a national hero for revealing God's laws. Evidence that Genesis must be reinterpreted to account for evidence that the earth was older than the traditional interpretation of six thousand years (an estimate based on the genealogies in the Old Testament) did not bother either Henslow or Sedgwick. One might interpret the word "days" in the Book of Genesis, for example, as figurative for periods, or posit a long gap between the creation of the heavens and earth and the six days of creation of animals, plants, and human beings. Indeed, thanks to the work of Sedgwick and others, by the mid-nineteenth century few Protestants (much less Catholics, who were used to non-literalist interpretations) interpreted Genesis as a literal account of the order and timing of creation. (The rise of six-day literal creationism would take a world war and be, at least initially, limited to the United States.)[43]

Confident that the Book of Scripture and the Book of Nature could not be in conflict, the extraordinary expansion of time demanded by recent discoveries in geology (examined in the next section) did not disturb these Cambridge naturalists. Indeed, Henslow reminded those who feared advances in science that although some ministers had rebuked and maligned geologists' conclusions, they were "now content to accept them as evidence of a Wisdom, Power, and Goodness, beyond any that former ignorance could ascribe to the works of that First Great Cause." Even, as we will see next, if naturalists proved that thousands of God's perfectly designed creatures had disappeared from the face of the earth.[44]

From Scarped Cliff and Quarried Stone

When Charles Darwin accepted the theory of independent creation, he did so in the face of new discoveries that, to some, called into question William Paley's entire worldview: thousands of animals and plants had been created only to become extinct. Neither William Paley nor Erasmus Darwin had had to wonder why, if God was so wise, powerful, and benevolent, so many plants and animals had gone extinct, leaving only fossilized vestiges of their presence. In 1800 no one knew entire flora and fauna had disappeared. In the midst of extolling the wisdom and goodness of God in the 1690s, John Ray, for example, had praised "the great Design of Providence to maintain and continue every Species." He cited the existence of males and females as evidence of that provision, combined with the "inexpugnable Appetite of Copulation" in both sexes. Ray was sure that amid "all the Endeavours and Contrivances of Man and Beast to destroy them, there is not to this Day one Species lost of such as are mention'd in Histories, and consequently and undoubtedly neither of such as were at first created." And if that was true, Ray concluded, God's close governance over his Creation was clear.[45]

Deists disagreed regarding the precise nature of God's governance but agreed that extinction must be impossible. In 1785 Thomas Jefferson summarized the turn-of-the-century consensus that none of God's creations had gone extinct as follows: "Such is the economy of nature, that no instance can be produced of her having permitted any race of her animals to become extinct; of her having formed any link in her great work so weak as to be broken."[46] Whether they emphasized God's benevolence, nature's, or a mix of the two, everyone was certain: species rarely, if ever, went extinct. Here is William Paley's version of this belief, composed in 1802: "Though there may be the appearance of failure in some of the details of Nature's works, in her great purposes there never are. Her species never fail. The provision which was originally made for continuing the replenishment of the world, has proved itself to be effectual through a long succession of ages." (Naturalists did not prove the dodo had gone extinct until the 1840s.)[47]

Then, around the turn of the century, Georges Cuvier (the French naturalist we met arguing against Lamarck's theory of evolution) drew upon the unparalleled collections of the National Museum of Natural History in Paris to prove that, in fact, entire floras and faunas had disappeared—repeatedly.[48]

The potential challenges the discovery of extinction posed for both natural theology and belief in a close personal God were most famously captured by the British poet Alfred Tennyson's poem *In Memoriam*, published in 1850 but composed as he grieved a beloved friend between 1833 and 1850. When nature seemed "so careless of the single life," bringing but one of fifty seeds to bear, then God and nature seemed "at strife." But surely one might stretch "lame hands of faith" and "faintly trust the larger hope," for nature was, at least, careful and protective of the "type," or species. Then nature, via the fossil record, replied: "'So careful of the type?' but no. From scarped cliff and quarried stone, She cries, 'A thousand types are gone: I care for nothing, all shall go.'"[49]

Why would a benevolent God create such marvelous, perfectly designed creatures, only to allow them to die out? This question had, of course, been asked before regarding the death of individuals, especially infants. In 1734 Benjamin Franklin had wondered why "so wise, so good and merciful a Creator . . . produce[d] *Myriads* of such exquisite Machines . . . but to be deposited in the dark Chambers of the Grave?" "Should," Franklin wrote, "an able and expert Artificer employ all his Time and his Skill in contriving and framing an exquisite Piece of *Clock-work*, which, when he had brought it to the utmost Perfection Wit and Art were capable of, and just set it a-going, he should suddenly dash it to pieces; would not every wise Man naturally infer, that his intense Application had disturb'd his Brain and impair'd his Reason?" (Franklin did not answer his questions by declaring God insane. Instead, he concluded that the extraordinary design evident in the bodies of infants proved heaven must exist, for otherwise high child mortality "can be in no wise consistent with the Justice and Wisdom of an infinite Being, to create to no end.")[50]

With the discovery of extinction, the scale of the mystery of why God created only to destroy had expanded to an almost unimaginable scale. Even deists, who had distanced God's governance from the details of the Creation, initially balked at the idea of extinction. Indeed, Cuvier's colleague Lamarck developed his theory of evolution in part because he could imagine no natural mechanism by which a species would be entirely destroyed, nor reconcile that destruction with belief in a benevolent Creator. Mastodons hadn't disappeared, Lamarck decided. They had transformed into elephants through an *upward law of development* (pushing organisms toward increasing complexity) and the *inheritance of acquired characteristics* (which explained variation, purposeful parts, and diversity).[51]

As we have seen, Lamarck's theory of evolution gained very little traction. Politically, scientifically, and theologically, it made little sense to most naturalists. And that meant that the challenge of extinction to the theory of special creation had to be faced: entire "perfectly designed" flora and fauna had disappeared from the face of the earth. Furthermore, the pattern as one moved from deep and presumably older deposits to more recent layers of the earth (often exposed through mining and excavation driven by the Industrial Revolution) posed profound additional mysteries. As organisms disappeared, any good comparative anatomist could tell that new species appeared in more recent layers. Where did they come from, and how did they appear? And how and why did *Homo sapiens* appear only in the most recent of these seemingly endless graves?

Cuvier provided answers that dominated natural history for the first half of the nineteenth century. Taking the observation that species are uniquely fit to particular environments as his starting point, Cuvier argued that extinction followed occasional catastrophic changes in the environment. Organisms composed of interconnected, functional parts would be driven to extinction by environmental change. Notice that, in Cuvier's hands, the discovery of extinction became an argument *against* evolution, for it seemed to prove Cuvier's claim that species could not evolve in the face of environmental change.

When Darwin arrived in Cambridge, Professor of Geology Adam Sedgwick had helped to develop a framework through which Scripture and the theory of special creation could be reconciled with the discovery of extinction. Like Henslow, Sedgwick believed that truths about nature could not be in conflict with Scripture. "Truths can never war against each other," Sedgwick wrote. "I affirm, therefore, that we have nothing to fear from the results of our inquiries, provided they be followed in the laborious, but secure road of honest induction" (i.e., carefully based on empirical evidence and the careful gathering of facts).[52]

Confident that two truths cannot contradict each other, Sedgwick taught students like Darwin how to map geological strata in the present and the fossil record in the past. During field trips, Darwin learned how natural processes (especially the movement of water) could transform landscapes. And he learned how perfect adaptation explained why animals and plants sometimes went extinct. Thus, even as governance via natural law became a hallmark of both science and faith, especially in geology, and even as the timeline expanded and the rhythm of creation (extinction-creation-extinction-creation) changed, the origin of species remained under the purview of independent acts of God and thus outside the realm of natural laws. Only "divine interposition," Sedgwick was certain, could explain purposeful parts.

In describing the appearance of new species in the fossil record as due to special interventions of God, Sedgwick updated the theory of special creation with the latest geological knowledge. As he mapped geological strata, he saw himself as documenting a history of independent, successive, progressive, new creations (by "progressive" Sedgwick did *not* mean species arose via evolution). Properly interpreted, he argued, the geological record mapped onto the order of creation described in Genesis, in which God prepared the earth for humanity: after all, human remains appeared only in the most recent of geological deposits.

Nature, God, and society were of a piece in the natural history and geology Darwin learned at Cambridge. In retirement, Sedgwick reflected on his life in science as follows: "I am thankful that I have spent so much of my life in direct communion with nature, which is the reflection of the power, wisdom, and goodness of God." He also preached to coal miners in Newcastle of the providential "economy of the coal-field" and their station relative to coal owners and capitalists. This might seem like a strict defense of the status quo, but politically, these Cambridge men believed in both progress and reform. Politically, both Henslow and Sedgwick were liberals, supporting limited electoral reform. Both were ardent abolitionists. Both argued for more support for the sciences within both universities (which tended to focus on the humanities) and the government. Throughout, they were confident, like most British naturalists (the word *scientists* would be coined by William Whewell in 1834), that the study of secondary causes revealed the nature of God's governance. They also believed, like John Ray and William Paley, that the evidence that species had been independently created would prevent the study of secondary causes from disconnecting God's governance from human lives and a stable British social order.[53]

However, it soon became clear that, despite Sedgwick's best efforts, the discovery of extinction could also be put to quite radical ends. Mary Shelley's last novel, *The Last Man* (published in 1826), though not read as often as *Frankenstein*, imagined a world in which a plague caused the extinction of human beings. (The cholera pandemic of 1817–1824 served as inspiration.) To anyone who believed that man was made in the image of God, closest to the angels, and that God designed all suffering for his benefit, Shelley imagined the unthinkable: the supposed pinnacle of creation, *Homo sapiens*, the one creature destined for heaven, might be annihilated just as the mastodons had been. Shelley proposed no heaven as compensation. She called the (presumed) capacity of humans to suffer more than animals (due to their ability to reflect, love, and form domestic attachments) a "defect, not an endowment." As scholar Lauren Cameron notes, Shelley imagined that "suffering is random, death happens without a greater cause, and nature acts without concern for individuals, much less species." This was a radical repudiation of Christianity's assumption that human beings were either under God's special care or worth so much more than lilies and swallows. But for some Shelley's vision made better sense of both human experience and the world.[54]

Adam Sedgwick took it upon himself to halt those who would use the discovery of extinction to either distance God's hand from creation or remove God altogether. Standing before the British Association for the Advancement of Science in 1844, he countered the "seductions of fanciful hypotheses" like transmutation by insisting that species appeared "not by the transmutation of those before existing—but by the repeated operation of creative power." "In his *ordinary dealings* with the natural world God works by second causes," Sedgwick explained, but in creating species he worked via the miraculous breathing of "a living spirit" into dead matter. Indeed, Sedgwick argued that the discovery of what he called "successive, progressive creations" provided an even stronger demonstration of God's close personal providence. After all, Sedgwick argued, a world in which new species appeared (via the miraculous breathing of a "living spirit") was not a world God had abandoned.[55]

The fact Sedgwick drew a direct connection between the special creation of animals and God's preservation of social hierarchies, the Anglican clergy, and the privileges of his profession and class did not go unnoticed. One of Britain's few public atheists, William Chilton, ridiculed Sedgwick's claim to be "God's 'own reporter.'" "How does he know this?" Chilton demanded, and where were the "*facts* which would warrant such a generalization?" But in the 1820s most naturalists, physicians, and anatomists agreed with Sedgwick that the facts *did* warrant such a generalization, and furthermore, that the generalization contained important lessons about the origin of affliction and path of salvation. On the more conservative side, the Anglican Professor of Medicine at Oxford John Kidd insisted "every sparrow which falls to the ground contains in its structure innumerable marks of the Divine care and kindness." The man who knows this "will be persuaded that every individual, however apparently humble and insignificant, will have his moral being dealt with according to the laws of God's wisdom and love." On the more liberal, Unitarian side, the physician and sanitation pioneer T. Southwood Smith appealed to the goodness of nature's laws as demonstration that, as "our reverend Master assures us . . . not even a swallow falleth to the ground, without the will of our heavenly Father; and that the very hairs of our head are all numbered; meaning, it is evident, that our most trifling concerns are appointed by him." Smith argued that, in general, this divine governance operated via secondary causes, but that those causes could only act in ways that God had appointed. Of course, a Unitarian version of this argument like Southwood Smith's might look non-Christian to Anglicans, but both sides argued that God's close personal providence could be upheld as knowledge of nature advanced.[56]

By background and temperament Darwin seemed set to join the ranks of this consensus regarding both the origin of species and God's close personal providence. When his studies for the ministry were interrupted in 1831 by an invitation to join the Admiralty ship HMS *Beagle* as a gentleman's companion and naturalist, his family, professors, and

Darwin himself thought it would be a temporary interruption. In 1832, at the age of twenty-three, he wrote from Rio de Janeiro to his sister Caroline: "Although I like this knocking about.—I find I steadily have a distant prospect of a very quiet parsonage, & I can see it even through a grove of Palms."[57]

The continued ties between natural theology (the study of nature) and revealed theology (the study of Scripture) can be seen in the quick response of Charles Darwin's uncle, Josiah Wedgwood, to the objections Darwin's father had put forth against his going on a voyage around the world. His father feared that the trip would be "disreputable" (after so much time with sailors) to Charles's career as a clergyman. He was also afraid Charles would not come home (a high percentage of naturalists on these ships died of yellow fever, malaria, typhus, cholera, or other diseases). But Wedgwood tried to assuage at least some of these fears. "I should not think that it would be in any degree disreputable to his character as a Clergyman," he wrote. "I should on the contrary think the offer honourable to him; and the pursuit of Natural History . . . is very suitable to a clergyman." That such a statement needed no further elaboration was thanks to William Paley, Darwin's professors, and generations of naturalists who saw their work as demonstrating the wisdom, power, and goodness of God. Darwin's father relented, and the *Beagle* obtained its final crew member.[58]

CHAPTER FOUR

Charles Darwin and Other People's Children

WITHIN A FEW MONTHS OF RETURNING HOME FROM HIS FIVE-YEAR voyage, Charles Darwin abandoned the theory of special creation. Prior to the voyage he had accepted the arguments of both William Paley and his Cambridge professors in full. Having changed his mind, sometime in 1838 Darwin privately wrote the following complaint in a small red notebook: "We can allow satellites, planets, suns, universe[s], nay whole systems of universe, of man, to be governed by laws, but the smallest insect, we wish to be created at once by special act, provided with its instincts its place in nature, its range."[1]

Darwin wrote these words of rebellion against the theory of special creation *after* his conversion to transmutation. But why did he adopt evolution as a better explanation of the origin of species in the first place? How did he decide that extending secondary causes to the question of the origin of species was the right and scientific thing to do? The myth of Darwin voyaging to the Galapagos Islands and immediately deciding that natural selection explained the extraordinary variation in the beaks of the island's finches is well known. Less well known is the fact that he did not even learn the relevant bird specimens were all finches until *after* he had returned home. During the voyage itself, dozens of puzzles that had very little to do with finch beaks crowded in upon him. All of these puzzles were inspired by his commitment to the theory of special creation and the closely associated belief in special providence.

Darwin suffered, and saw his fellow crewmen suffer, during his five years traveling the globe. He witnessed crew members die, and helped rescue castaways close to death. He almost died twice—once almost frozen on a beach, once of a fever. But it was the suffering of those outside his own experience, including the suffering of children, that posed the most profound mysteries for someone who believed in both humanity's status as God's greatest creation and God's close personal governance. If it wasn't for these puzzles, he may never have wondered about finch beaks, or become unsatisfied with explaining "the smallest insect" via a special act.

What Would the Disbeliever Say to This?

At twenty-two years old, Darwin walked up the plank of the HMS *Beagle*, a firm believer in the theory of special creation and the biblical promise of heaven. He later wrote that "Whilst on board the *Beagle* I was quite orthodox, and I remember being heartily laughed at by several of the officers (though themselves orthodox) for quoting the Bible as an unanswerable authority on some point of morality."[2]

His head was full of Paleyan assumptions when he left England: that organisms were designed by a benevolent personal God and that human beings were created in God's image. As an abolitionist, he also believed that all human beings were the same species and thus no human being should be the slave of another. He believed in special providence, and that on balance the world was a happy place. And he agreed with the methodological assumptions of Professors Henslow and Sedgwick, two of the most influential naturalists in Britain: First, that appeal to God (the first cause) rather than natural mechanisms (secondary causes) was appropriate for explaining the ultimate origin of species (and life itself); and second, that explaining phenomena *post-creation* (including generation, development, and reproduction) must be done via the purely mechanistic, naturalistic terms of secondary causation. Both of these assumptions influenced what he expected to see in South America and his responses to what he did see.

Given his time at medical school in Edinburgh, Darwin knew a great deal about alternative visions of both God and nature. He had learned about theories of "transmutation" that held species evolved from other species, men arose from apes, and material forces created everything without divine guidance. He had read his grandfather Erasmus Darwin's book *Zoonomia* and walked with Robert Grant as the latter praised Lamarck's ideas. He had been present at club meetings where members discussed materialist explanations of the human mind (and then promptly expunged the conversations from the minutes). But he had no reason to question the theory of special creation taught in both the established medical schools and by his professors at Cambridge. And he had a great many reasons—political, social, theological, and scientific—to dismiss transmutation as an impious, unscientific, and even seditious explanation of the world.

The voyage that eventually inspired Darwin to change his mind was supposed to last two years. Captain Fitzroy's task was to expand British political and economic influence in South America by mapping the continent's southwestern coastline. Fitzroy was fastidious, to put it mildly, and the two-year journey eventually stretched into five. Wracked with seasickness when onboard, Darwin spent two-thirds of the voyage on land collecting natural history specimens and observing the continent's geology and people while Fitzroy sailed up and down the coast to make sure his maps would guide British ships safely and efficiently.

Inevitably, Darwin's belief in the theory of special creation colored everything he saw. Observing the conical pit of an Australian "Lion Ant" similar to those he knew in England yet thousands of miles away, he wrote: "Now what would the disbeliever say to this? Would any two workmen ever hit on so beautiful, so simple & yet so artificial a contrivance? It cannot be thought so. The same hand has surely worked throughout the universe." Standing beneath the canopy of the rainforest in Brazil, a temple "filled with the varied productions of the God of Nature," he wrote, he was certain "that no one can stand in these solitudes unmoved, and not feel that there is more in man than the mere breath of his body." Professors Henslow and Sedgwick would have been proud.[3]

Darwin did rebel against his devout Cambridge professors on one major point, however. During the first months of the voyage he became an ardent disciple of a lawyer-turned-geologist named Charles Lyell.

Lyell had launched an aggressive campaign for a new set of methodological assumptions that he believed would provide a better guide to the geological past than those Darwin had learned from Adam Sedgwick. Both Sedgwick and Georges Cuvier held that one could posit processes greater in degree and different in kind *if* they explained the patterns observed. By contrast, Lyell argued that geologists must use *only processes uniform (i.e., the same) in kind and degree as those observed in the present* to explain geological formations. Drawing on this methodological rule (eventually called uniformitarianism), Lyell extended the age of the earth beyond what even Sedgwick was willing to countenance. After all, if one must explain mountains and canyons by appealing solely to the rates of deposition and erosion observed in the present, then these processes were so slow that canyons and mountain ranges must have taken an unimaginably long time to form.

Lyell also applied uniformitarianism to the problem of extinction. When citing Malthus's laws of population, Paley had consoled himself with the belief that the limits "fixed by nature" had not yet been reached in any country. Charles Lyell, writing a generation later and after the discovery of extinction, begged to differ. The limits "fixed by nature" had been reached again and again. Unwilling to imagine widespread catastrophes (like the continental floods posited by some naturalists) to explain extinction, Lyell turned Malthus's struggle for existence into the causal mechanism for extinction. "In the universal struggle for existence," Lyell wrote in *Principles of Geology*, "the right of the strongest eventually prevails," and the strength and durability of a species depended on its "prolificness." Lyell did *not* use the "struggle for existence" to explain the *creation* of species. The latter he attributed, like most naturalists, to God.[4]

Before they left England, Captain Fitzroy had given Darwin a copy of the first volume of Lyell's three-volume *Principles of Geology*. Professor Henslow had recommended Darwin procure a copy for the voyage, but warned him "on no account to accept the views therein advocated." As Darwin read *Principles* and applied its explanatory framework to the landscapes of the *Beagle*'s first stop, the Cape Verde Islands, he ignored Henslow's advice and adopted uniformitarian methods completely. He thus became Lyell's "first and, at that time, only scientific disciple."[5]

Having adopted Lyell's uniformitarian methods, Darwin also embraced the slow "struggle for existence" (rather than Cuvier and Sedgwick's rapid "catastrophes") to explain extinction. Like Lyell, he did not extend either the struggle for existence or uniformitarian methods to the *creation* of species. Indeed, Darwin was entirely convinced by Lyell's careful argument against Lamarck's theory of transmutation (included in volume II of *Principles of Geology*, which Darwin obtained when the *Beagle* picked up the mail in Montevideo, Brazil, in November 1832). Well aware of how his vision of an unimaginably old, constantly changing Earth might be used by transmutationists, Lyell gave a strictly empirical assessment of Lamarck's views. "No positive fact is cited," Lyell argued, "to exemplify the substitution of some *entirely new* sense, faculty, or organ, in the room of some other suppressed as useless." All was fancy and speculation, Lyell argued, fiction rather than fact.[6]

Lyell explained that the primary appeal of Lamarck's theory was its rejection of "the repeated intervention of a First Cause" in the creation and extinction of species. Indeed, as a Unitarian Christian, Lyell himself was sympathetic to the search for secondary causes to explain phenomena. He argued, for example, that purely secondary causes could explain both extinction and geological landscapes. But he insisted that Lamarck's theory of transmutation did not add up on scientific grounds. His secondary cause, or mechanism of species change, was unconvincing and speculative, and observation demonstrated that species were constant. Even the most variable domesticated forms, such as dogs, never crossed the species boundary. No new organs had been produced by domestic breeding. The oldest specimens of animals and plants, those mummified in Egyptian tombs, remained exactly the same as those in the present. And that was as far back as one might go and still work within the rules of science.

For much of the voyage Darwin found Lyell's explanation of the origin and extinction of species as acceptable as his geology. The earth was very old, landscapes had changed a great deal, and animals and plants could not "adapt" *themselves* to those changes. Indeed, the extinct giant ground sloths, giant armadillos, and other fossils he and the *Beagle* crew dug out of riverbanks in eastern South America seemed to prove as much.

Darwin now had a trio of teachers influencing how he saw the world: Adam Sedgwick, John Henslow, and Charles Lyell. Despite their theological differences (Sedgwick and Henslow were Anglican, Lyell Unitarian), each of these mentors firmly believed in something called the "dignity of man." That God created human beings "in His image," with the divinely endowed capacity to both reason and know good from evil, was *the* nonnegotiable point at which all geologists, naturalists, and anatomists' talk of secondary causes, natural laws, and mechanisms halted. Indeed, Lyell argued against Sedgwick's theory of "successive, progressive creations" because he feared transmutationists might use Sedgwick's description of the fossil record as progressive to break the tight barrier between human beings and animals. Portraying the fossil record as a progression from simple to complex organisms (even as independent creations), Lyell warned, made it easier for transmutationists to argue that humans had evolved via purely naturalistic, mechanistic means from other animals. (Lyell insisted the fossil record was too poorly preserved to argue mammals had not existed since the beginning of time, a stance that Darwin would later turn to evolutionists' advantage.)[7]

Most of Lyell's anti-evolution chapter in *Principles of Geology* centered on empirical evidence against transmutation. But he could not hide his disdain as he described Lamarck's theory of "the transformation of the Orang-Outang into the human species." In "the last grand step in the progressive scheme," Lyell wrote, "the orang-outang, having been already evolved out of a monad, is made slowly to attain the attributes and dignity of man." Historian Janet Browne notes that Lyell was "generally aghast at the thought of animal ancestry" and wished to defend "an unbridgeable divide between mankind and the beasts." Given that the "dignity of man" (the unique, divinely conferred rational and moral attributes of human beings) is where Lyell halted as he used secondary causes to explain the past, attending to Darwin's changing attitudes toward his own kind is crucial to understanding why, eventually, Darwin refused to halt. We must understand why Darwin decided that first, he and his fellows were not, in fact, so much better than swallows and lilies, and second, God did not seem to be taking special care of human beings, much less falling sparrows.[8]

THE DIGNITY OF MAN

The Darwin-Wedgwood family's abolitionist background is critical to understanding how Darwin's ideas about human beings changed over the course of the voyage. Darwin's grandfathers Erasmus Darwin and Josiah Wedgwood had famously asked—the one on a jasperware medallion, the other in a poem—on behalf of those enslaved: "Am I not a Man and a Brother?" Their consistent financial support helped the generation-long campaign to pass the Slave Trade Act of 1807, which ended the legal trade in human beings on British ships. Charles's grandfather Josiah Wedgwood financed the abolitionist Thomas Clarkson, and his uncle (and later father-in-law) Josiah Wedgwood was a member of the London abolition committee, financed the Hanley and Shelton Anti-Slavery Society, and printed propaganda tracts. Darwin's aunt Sarah donated tens of thousands of pounds to anti-slavery societies. Uncle Josiah was in parliament as an abolitionist when the Emancipation Act became law in August 1833 and legal slavery ended in the British Empire.[9]

During the voyage, Darwin's sisters kept him apprised of that long-awaited event, and its disappointing flaws, as he traveled in South America. The act did not free enslaved men and women immediately, in contrast to the hopes of "immediatist" abolitionists like members of the Darwin family. Instead, it established an "apprenticeship" system, and even that was delayed for months. The Darwin and Wedgwood clan were also disappointed by the fact Parliament had voted to provide compensation to planters for their "lost property." In October Susan wrote to Darwin: "You will rejoice as much as we do over Slavery being abolished, but it is a pity the Apprenticeship does not commence till next August as that is a great while for the poor Slaves to be at the mercy of the Planters who I sh[d]. think w[d]. treat them worse than ever.—I grudge too very much the 20 million compensation money: but perhaps it would never have been settled without this sum."[10]

Both during and after the voyage, Charles Darwin assumed that the answer to his grandfathers' supplication on behalf of enslaved humanity must be, according to both nature and Scripture, a resounding "Yes." Originally rooted in Scripture, this assumption had been backed up by personal experience. While avoiding the surgery theatre in Edinburgh, he had spent hours learning taxidermy (how to stuff and preserve dead animals) from a man named John Edmonstone. Born into slavery (in then-British Guiana), Edmonstone had travelled to England with the explorer Charles Waterton. There, he secured his freedom and took a job at the university's natural history museum. He "gained his livelihood by stuffing birds," Darwin recalled later, "which he did excellently: he gave me lessons for payment, and I often used to sit with him, for he was a very pleasant and intelligent man." Darwin would remember Edmonstone when, traveling in Brazil or at home, he heard slavery's defenders claim that (as his friend Reverend John Innes wrote four decades later) God had created some men "who must be made to work and better men able to make them." (Darwin replied: "I consider myself a good way ahead of you, as far as this goes.")[11]

Charles Darwin firmly believed that, amid variation, all people were like him with respect to any trait that mattered for determining another human being's moral status. By today's standards, his humanitarianism seems of a limited kind. It did not equate to egalitarianism, and he was quite willing to rank groups as "higher" or "lower" depending on how similar they seemed to Englishmen. (White abolitionism rarely equated to egalitarianism, as evident in some American abolitionists' insistence that freed men and women should be "colonized" back to Africa to avoid an otherwise inevitable "struggle for existence" between the "races.") Yet Darwin's assumption that all human beings belonged to the same species meant he believed that he could imagine, at the most fundamental, physiological level, what another human being was feeling. This assumption would be crucial to the stance he eventually took on the meaning of human suffering, as he traveled through a country in which human beings enslaved and sold other human beings, including children.

A few months before the Slavery Abolition Act, Darwin wrote from Brazil to his sister Catherine of "what a proud thing for England" it would be if she was the first nation to utterly abolish slavery. (He seems to have forgotten about the Haitian Revolution completely while writing this sentence, though he remembered it a few lines later.) "I was told before leaving England," he continued, "that after living in Slave countries: all my opinions would be altered; the only alteration I am aware of is forming a much higher estimate of the negro character." It was hard, he wrote, not to wish for Brazil to follow the example of the revolution, led entirely by enslaved people, in Haiti. He was certain (in contrast, as we will see, to some of his fellow naturalists) that those of African descent were perfectly capable of success should such a "wonderful" event occur. He concluded his letter by urging that his sister contact "some of the Anti-Slavery people" to investigate a man in Rio who, though paid to prevent the landing of slave ships, was in fact facilitating the smuggling of men, women, and children from Africa into the Brazilian trade.[12]

Children and the ties of family figured prominently in Darwin's accounts of what he witnessed in Brazil. Darwin did not have his own children at this point. But he had a good imagination (he would later call imagination a foundation of the sentiment of sympathy) and believed that other human beings suffered and felt pain in the same way he did. That belief was initially rooted in biblical assumptions about the common origin of all human beings, or *monogenism* (of one origin). Both Unitarians and Anglicans, the contrasting heritages of Christianity in which Darwin grew up, appealed to passages in Genesis and Acts to insist that "God hath made of one blood all the nations of men" and that *all* human beings were made in the image of God.

As a medical student, Darwin learned about this unity in physiological terms in addition to biblical ones, but the lesson was supposed to be the same: one human being could always imagine what another human being's pain felt like. When Darwin stayed in a house where "a young household mulatto, daily and hourly, was reviled, beaten, and persecuted enough to break the spirit of the lowest animal" or witnessed "a little boy, six or seven years old, struck thrice with a horse-whip (before I could interfere) on his naked head, for having handed me a glass of water not quite clean," he never doubted that he could imagine the pain these children felt because they were, in all essentials that mattered, in the same taxonomic category as Europeans.[13]

By contrast, slavery's defenders denied human unity, placed non-Europeans in separate taxonomic categories, and then insisted some "races" felt pain *less*. Physicians and naturalists who called themselves "pluralists" (and later polygenists) argued that each of the major human lineages (sometimes five, sometimes seven, sometimes—as Darwin later wrote with disdain—twenty-seven) had either been independently created by God, or evolved from different primates. The seventeen-volume *Dictionnaire Classique d'Histoire Naturelle* (which the *Beagle* carried in its library throughout the voyage) divided humans into fifteen different *species*. In other words, some naturalists classified certain human beings outside the category of *Homo sapiens*. In doing so they pushed non-Europeans outside the boundaries of the Golden Rule ("do unto others as you would have done unto you") altogether.[14]

The implications of these hierarchical classifications, whether within the boundaries of *Homo sapiens* or not, were profound. Having classified Africans and Europeans as different races based on presumed physical differences, Thomas Jefferson insisted in his 1785 *Notes on Virginia* that the "griefs" of the former "are transient." The African race, he wrote, felt afflictions and suffering that "render it doubtful whether Heaven has given life to us in mercy or in wrath" *less*, and forgot them sooner. Thus, notes scholar Heather Andrea Williams, "Jefferson suggested that whereas in the face of a tragic loss a white person would question why God had placed him on earth, this question would not arise for black people because they felt pain superficially and only briefly."[15]

The view that those of African descent, including children, did not have the same capacity to feel pain, suffering, and grief proved a common trope in the pro-slavery camp. Some doctors in the US South even claimed Africans carried a hereditary disease called "dysaesthesia aethiopica" that made them less able to sense pain. Such claims not only justified violence, but because the ability to suffer and feel pain was tightly related to sentience, they called into question Africans' status as human beings.[16]

Enslaved or formerly enslaved men and women with access to printing presses fought back against these claims when they could. In 1773 Phillis Wheatley had demanded that readers imagine her parents' capacity to suffer as equal to their own:

What pangs excruciating must molest,
What sorrows labour in my parent's breasts

Steel'd was that soul and by no misery mov'd
That from a father seiz'd his babe belov'd.[17]

A little over fifty years later, in 1831 (the year Darwin left on the voyage), the first autobiographical account by a formerly enslaved woman, Mary Prince, tried to make the point again. Prince told her story, she explained, so that "good people in England might hear from a slave what a slave felt and suffered." Her account of "our mother, weeping as she went," forced to take her own children to the market where they would be sold, is a tale of child loss that had nothing to do with disease or accident, but everything to do with what one human being was willing to do to another. "My heart throbbed with grief and terror so violently," Prince wrote, "that I pressed my hands quite tightly across my breast, but I could not keep it still, and it continued to leap as though it would burst out of my body." No one near seemed able to imagine that the pain that wrung her and her mother's hearts was like their own. "Slavery hardens white people's hearts towards the blacks," she explained, "and many of them were not slow to make their remarks upon us aloud, without regard to our grief—though their light words fell like cayenne on the fresh wounds of our hearts. Oh those white people have small hearts who can only feel for themselves."[18]

As an abolitionist, Darwin witnessed Brazil's system of slavery completely certain that all human beings felt the same kind of emotional, mental, and physical pain as himself, and that he could therefore imagine what they might be feeling. Amid all the suffering he witnessed during the voyage, he felt the forced separation of parents and children most keenly. "I was present," he later wrote, "when a kind-hearted man was on the point of separating forever the men, women, and little children of a large number of families who had long lived together."[19] In his published account of the voyage, Darwin's description of what he had witnessed was driven by the belief that one could and should try to imagine what the enslaved human beings suffered:

Those who look tenderly at the slave-owner, and with a cold heart at the slave, never seem to put themselves into the position of the

latter;—what a cheerless prospect, with not even a hope of change!
Picture to yourself the chance, ever hanging over you, of your wife
and your little children—those objects which nature urges even the
slave to call his own—being torn from you and sold like beasts to the
first bidder![20]

Darwin's abolitionist loyalties had almost ended his voyage just a few
months after it began. He was indignant and disgusted when he heard
men offering excuses for slavery. When Captain Fitzroy coolly trusted
enslaved men's "no" when a Brazilian "master" asked them whether they
would prefer to be free, Darwin asked Fitzroy whether he thought men
in slavery could answer such a question truthfully. Fitzroy responded by
kicking him out of the captain's quarters, and for twenty-four hours Darwin thought he would be returning to England on the next ship home.
Fitzroy cooled off by morning, and they avoided the subject for the rest
of the voyage.[21]

When he was home and safely out of Fitzroy's control, Darwin let
loose in the pages of his contribution to *The Narrative of the Voyages of*
His Majesty's Ships Adventure and Beagle. He described how, near Rio de
Janeiro, he lived opposite an old lady who kept screws to crush the fingers
of the enslaved women in her home. He had seen a grown man "tremble
at a mere glance from his master's eye." He was overwhelmed by shame
when a man ferrying him across a river took his gestures indicating direction as an attempt to strike the man. "He, I suppose, thought I was in a
passion," Darwin wrote, "and was going to strike him; for instantly, with
a frightened look and half-shut eyes, he dropped his hands. I shall never
forget my feelings of surprise, disgust, and shame, at seeing a great powerful man afraid even to ward off a blow, directed, as he thought, at his
face. This man had been trained to a degradation lower than the slavery of
the most helpless animal." Darwin then defended mentioning "the above
revolting details" because he had met several people who spoke of slavery
as a tolerable evil. "Such enquirers will ask slaves about their condition;
they forget that the slave must indeed be dull who does not calculate on
the chance of his answer reaching his master's ears." He had made the

point that had so offended Fitzroy public, in a volume with "under the command of Captain Fitzroy" on its title page.[22]

Note that, in contrast to many of his contemporaries, Darwin wrote of the institution of slavery, not nature, as having "trained" men to "degradation." In doing so he implied that any presumed differences between Europeans and Africans under conditions of slavery were due to their environment. Of course, hierarchies based on "nurture" (rather than nature) could be just as racist in their assumptions as those rooted in nature. But Darwin's emphasis on environment and education as "degrading" or "raising" human beings up the ladder of moral and intellectual progress was, in its assumption that any human being had the capacity to move up or down, very different from polygenist views. Polygenists, whether they decided racial differences arose from special creation or evolution or some complex mix of the two, insisted human variation reflected essential, natural, and therefore unchangeable, boundaries.

In a world in which most applied the term *savage* to non-Europeans, Darwin wrote of plantation owners of Spanish descent as "savage masters" and called his countrymen who defended slavery "polished savages." As we will see, years later, intent on breaking the boundary between humans and animals by demonstrating the existence of gradations within *Homo sapiens*, Darwin would confidently write of "obvious" distinctions in intellectual and moral capacity between human "races." The *one* capacity he never wrote of as variable was the capacity to suffer both pain and grief. He always insisted that all human beings suffered alike. And because Darwin believed that all humans had the same capacity to suffer, his visit to a slave country had a profound influence on the questions he eventually launched at Paley's explanation of the world and, ultimately, God. Why did a benevolent God allow such things? What possible "benevolent purpose" could exist behind God's permission of such a horrible institution? Yes, God's endowment of free will meant humanity erred, but how could human beings treat each other like this? Taught to believe in a divinely conferred "dignity of man," Darwin's loss of faith in his own kind influenced his ability to eventually question the assumptions that drove both Paley's and his professors' explanations of the world. In recounting how slavery separated children from parents, he could locate no justification, no benevolence, and no wisdom: "These deeds are done and palliated by men who profess to love their neighbours as themselves, who believe in God, and pray that His Will be done on earth!" he cried. "It makes one's blood boil, yet heart tremble, to think that we Englishmen and our American descendants, with their boastful cry of liberty, have been and are so guilty."[23]

Charles Lyell had feared what transmutation would do to belief in the "dignity of man." Faced with how human beings treated other human beings, Darwin began wondering whether a divinely endowed "dignity of man" even existed. And once that doubt entered his mind, the tight barrier (insisted upon by Lyell, Sedgwick, and Henslow) between humans and animals began to crumble. After he left Brazil, his experience with Indigenous peoples in the southern regions of South America and the South Pacific slowly but steadily crumbled the boundary to the ground.

WITH DIFFICULTY WE SEE A FELLOW CREATURE

As a naturalist on an Admiralty ship, Darwin traveled the routes of imperial expansion, a firm believer in the good of spreading Christianity, capitalism, and civilization. (As we will see, these three things went together in Darwin's eyes as the best means of progress, *even after* he abandoned belief in Christian doctrines.) Meanwhile, tormented with seasickness, Darwin was often approached during the *Beagle's* first passage from England to South America by a fourteen-year-old passenger from the Tierra del Fuego, Jemmy Button. "Poor, poor fellow!" Jemmy would say. Composing his account of the voyage upon his return home, Charles remembered Jemmy's kind words, noting "he was remarkably sympathetic with any one in pain."[24]

As historian Janet Browne notes, "European cultural chauvinism of the most obvious kind" was behind Jemmy's presence on the HMS *Beagle*. During his previous voyage to the region, Captain Fitzroy had basically kidnapped Jemmy (of the Yahgan people) and three other individuals of a nearby group (Elleparu, Yokcushly, and a man the British called "Boat Memory" but whose given name is unrecorded). He took the four all the way to England with the intent of returning them to Tierra del Fuego in a few years and founding an Anglican mission. Swept to England on a wave of British imperialism, all four Fuegians learned English, how to wear European clothing, and how to bow and curtsy before King William IV and Queen Adelaide. Though vaccinated against smallpox upon arrival in England and kept by Captain Fitzroy in relative isolation to avoid disease, "Boat Memory" died of smallpox before he could be, as planned, returned home to help "civilize" and "Christianize" his family and friends.[25]

The story of the Fuegians on the HMS *Beagle* is emblematic of a self-righteous paternalism that characterized the entire British imperial endeavor. Fitzroy clearly believed himself more humane and Christian than General Rosas, whom the *Beagle* crew met in the midst of the general's brutal campaign to actively exterminate Indigenous communities in Argentina. He thought he was saving the Fuegians on both Earth and in heaven (by his own standards and terms, of course). In contrast to Rosas and many other settler-colonials, Fitzroy was absolutely certain that Fuegians were human beings, and he firmly believed that they deserved earthly and eternal salvation as much as any Englishman.

Both Fitzroy and Darwin saw Jemmy's people as lower on a "ladder of progress" that Europeans got to define, but they also believed in all human beings' God-given capacity to advance up that ladder via Christianity, education, and the adoption of private property. Both supported missionaries under the assumption that Christianity moved people up the ladder of progress while saving souls. Intent on justifying his actions as benevolent, Fitzroy urged skeptics of such "philanthropy" to remember that "unwilling as we may be to consider ourselves even remotely descended from human beings in such a state, the reflection that Cæsar found the Britons painted and clothed in skins, like these Fuegians, cannot fail to augment an interest excited by their childish ignorance of matters familiar to civilized man, and by their healthy, independent state of existence." Like Fitzroy, Darwin knew that noting Jemmy Button's remarkable ability to sympathize with anyone in pain was worth including because plenty of his readers did not automatically assume a boy of Jemmy's "kind" could feel the same things an Englishman could.[26]

We know a lot about what Darwin and Fitzroy thought about Jemmy. We do not know about Jemmy from his own words. No one taught him how to write. Instead they wrote about him. Describing the passage from England to South America, for example, Darwin noted he had a "nice disposition," was merry, and often laughed.[27]

As a monogenist, Darwin was clearly fascinated by the differences he could find between his and Jemmy's view of the world. He was bemused by Jemmy's insistence that there was no Devil in his land, and by Yokcushly's explanation of affliction (the latter warned the crew of much rain and snow when one of them killed some very young ducklings for specimens, "evidently a retributive punishment for wasting human food"). The Yahgan did not even seem to have a religion in the sense that Englishmen understood the term. "Captain Fitz Roy could never ascertain that the Fuegians have any distinct belief in a future life," Darwin wrote. "We have no reason to believe that they perform any sort of religious worship." It is important to note that, since the seventeenth century, some British philosophers had claimed that belief in God was a uniquely human trait that separated human beings from the rest of the animal world. As historian Matthew Day has noted, this distinction allowed the agents of colonial expansion to push Indigenous populations (who did not seem religious by European criteria) *outside* the boundary of *Homo sapiens*. Darwin refused, both in the 1830s and later, to do so. And that refusal made what he witnessed all the more intriguing.[28]

Jemmy taught him, for example, about the great differences in ritual, belief, and social rules that existed between cultures. Jemmy had been around twelve when Fitzroy took him away. Homesick himself, Darwin witnessed the boy's reunion with his family three years later, and struggled to understand Jemmy's behavior based on his own experience of family and domestic life. "Me no help it," Jemmy shrugged when he learned of the death of his father. Darwin was mystified by the lack of demonstrative affection between mother, brother, and son when they were reunited. "The meeting was less interesting," he recorded, "than that between a horse, turned out into a field, when he joins an old companion." They had simply stared at each other, and the mother quickly left to look after her canoe. But in his account of the voyage he was careful to point out that it was the *display* of grief that varied, rather than grief itself. "We heard, however," he explained, "through York Minster (Elleparu) that the mother had been inconsolable for the loss of Jemmy, and had searched everywhere for him, thinking that he might have been left after having been taken in the boat."[29]

It is within the context of Darwin's strong commitment to the monogenist unity of man and his fundamental assumption that Jemmy was a fellow human being (rather than of another species), that one must read Darwin's shocked account (written to his sister Caroline) of seeing Jemmy's relatives for the first time:

> *We here saw the native Fuegian; an untamed savage is I really think one of the most extraordinary spectacles in the world.—the difference between a domesticated & wild animal is far more strikingly marked in man.—in the naked barbarian, with his body coated with paint, whose very gestures, whether they may be peaceable or hostile are unintelligible, with difficulty we see a fellow-creature.[30]*

The crucial point here is that, while it may have been difficult, Darwin did see a fellow creature. He knew, based upon Scripture, comparative anatomy, and his time with Jemmy, that Jemmy's relatives were fully human beings. He had been struck by how similar Jemmy and the other Fuegians on the *Beagle* were in intellect and disposition to Englishmen. Later, when revising his letters home for his *Journal of Researches*, Darwin added of Jemmy in particular: "It seems yet wonderful to me, when I think over all his many good qualities, that he should have been of the same race, and doubtless partaken of the same character, with the miserable, degraded savages whom we first met here." The difference between "civilized" Jemmy and his "uncivilized" relatives was at times overwhelming. Here is Darwin's description of individuals in a canoe that pulled alongside the boat as they came to the shore of Wollaston Island, "the most abject and miserable creatures I anywhere beheld":

> *These Fuegians in the canoe were quite naked, and even one full-grown woman was absolutely so. It was raining heavily, and the fresh water, together with the spray, trickled down her body. In another harbour not far distant, a woman, who was suckling a*

recently-born child, came one day alongside the vessel, and remained there out of mere curiosity, whilst the sleet fell and thawed on her naked bosom, and on the skin of her naked baby! These poor wretches were stunted in their growth, their hideous faces bedaubed with white paint, their skins filthy and greasy, their hair entangled, their voices discordant, and their gestures violent. Viewing such men, one can hardly make one's self believe that they are fellow-creatures, and inhabitants of the same world.[31]

And yet Darwin firmly believed the Fuegians *were* fellow creatures. "It was this rejection of separate origins," notes historian Janet Browne, "that made the whole experience of Tierra del Fuego so painfully interesting to him."[32] He was struck, for example, by how quickly Jemmy abandoned his "civilized" ways. Fitzroy's plan had always been that Jemmy, Elleparu, and Yokcushly would see the virtue of British ways and beliefs, and set to "civilizing" their homeland. When the *Beagle* visited Jemmy again a year later, Darwin was just as shocked at how quickly he had returned to his old ways:

We could hardly recognize poor Jemmy; instead of the clean, well-dressed stout lad we left him, we found him a naked thin squalid savage. York & Fuegia had moved to their own country some months ago; the former having stolen all Jemmy's clothes: Now he had nothing, excepting a bit of blanket round his waist.—Poor Jemmy was very glad to see us & with his usual good feeling brought several presents (otter skins which are most valuable to themselves) for his old friends.—The Captain offered to take him to England, but this, to our surprise, he at once refused: in the evening his young wife came alongside & showed us the reason: He was quite contented; last year in the height of his indignation, he said "his country people no sabe nothing.—damned fools" now they were very good people, with too much to eat & all the luxuries of life.[33]

Darwin clearly found Jemmy's decision impossible to understand. In his account of the voyage, he had nothing good to say about either the countryside or the life Jemmy's people lived. "The country is a broken mass of wild rocks," he wrote, "lofty hills, and useless forests: and these are viewed through mists and endless storms." With little habitable land, the people had to wander ceaselessly, traveling in their "wretched canoes." Homesick for his comfortable family life and beloved sisters and father, he confidently declared that surely the Fuegians "cannot know the feeling of having a home, and still less that of domestic affection; for the husband is to the wife a brutal master to a laborious slave." He then repeated an old story originally told by John Byron, who had traveled the region in the 1740s: "Was a more horrid deed ever perpetrated, than that witnessed on the west coast by Byron, who saw a wretched mother pick up her bleeding dying infant-boy, whom her husband had mercilessly dashed on the stones for dropping a basket of sea-eggs!"[34]

Darwin thought much on how far Jemmy had come in a few short years. He then made the imaginative leap of extending his observation of the extraordinary malleability of human nature into the history of his own family, the British, and, ultimately, long after he had sailed away from Fuegian shores, *Homo sapiens* itself. Thanks to his friendship with Jemmy, his observations of slavery, and his belief in the "unity of man," Darwin pondered how tenuous the barrier between "savagery and civilization" really was. He began thinking of "savage" as a description of certain behaviors rather than, as some of his contemporaries did, a category into which all non-Europeans could be conveniently slotted. What Darwin learned from the difference between young Jemmy (after a short time in England) and his Fuegian relatives was how quickly "civilization" could be both conferred and removed from *anyone*. Witnessing stranded sailors and his own feelings amid deprivation taught him how easily the "civilized" man could be made "savage," himself and all other Britons included. The so-called "dignity of man," so prized by Lyell, might be both *taught* and *unlearned*, rather than divinely conferred.[35]

Upon their final farewell, Darwin wrote: "Every soul on board was sorry to shake hands with poor Jemmy for the last time, as we were glad to have seen him. I hope and have little doubt he will be as happy as if he had never left his country which is more than I formerly thought." Years later, he exchanged letters with those trying to arrange support for Jemmy's orphaned grandsons. As he enclosed a one-pound-per-year subscription, Darwin expressed the hope that his correspondent had carefully considered whether it would be a real kindness to educate one of the boys. The comment is evidence of some lingering doubt that Fitzroy—and the entire British business of spreading "civilization"—had been a "real kindness."[36]

By the time he wrote this letter in 1878, Darwin would have known that epidemics had by that time decimated the Yahgans (a smallpox epidemic in 1876 hit particularly hard), and Jemmy and his own children were dead, having succumbed to the mysterious high mortality that tracked the steps of Europeans.

WHEREVER THE EUROPEAN HAS TROD

Like Charles Lyell, Adam Sedgwick, John Henslow, and William Paley, Darwin began the voyage certain of a firm, creative break between human beings and other creatures. Humanity's possession of both a moral and rational soul, endowed by God, was supposed to separate *Homo sapiens* qualitatively from the rest of creation. As his observations of his own kind slowly chipped away at this assumption, he was also, of course, observing nonhuman creatures and the landscape upon which they lived. Over the course of the five-year voyage, he amassed a series of puzzles and mysteries in the patterns he observed that would eventually be crucial to his willingness to embrace an evolutionary explanation of the origin of species.

First, there were puzzles in the geographical and chronological distribution of fossilized forms. For almost a generation, naturalists had known that entire flora and fauna had disappeared from the face of the earth. But explanations for that extinction, as we have seen, varied. Cuvier and Sedgwick explained extinction as the result of large-scale catastrophes that changed organisms' "conditions of existence." Having been designed for one set of conditions, organisms would no longer fit the suddenly changed conditions and would die out. By contrast, having repudiated change via catastrophes, Lyell explained extinction as the result of a slower, "uniformitarian" process: a struggle for existence that resulted in competition between species and caused the "weaker" forms to die out. (Recall that Lyell did not posit that struggle as a means by which new species might be formed.)

While traveling the coast of South America, Darwin fully adopted Lyell's explanation of the fossils the *Beagle* crew unearthed on the beaches: a giant ground sloth (*Megatherium*), a giant armadillo (*Glyptodon*), and other mammals. But Darwin did wonder about the patterns evident when he compared these extinct forms with living South American animals. He unearthed giant ground sloths in the same region that much smaller, tree-bound, two- and three-toed sloths now roamed, and fossils of giant armidillos in the same region as much smaller armadillos now lived. Committed to the theory of special creation, which presumed divine purpose in all facts, Darwin pondered: Why had God allowed giant ground sloths and giant armadillos to go extinct, yet created new forms within the same family of animals in their place? Why not replace extinct species with something completely different? Why recreate on the same "armadillo" and "sloth" plan? These patterns in fossilized forms and the living inhabitants of particular geographical regions struck Darwin as an interesting puzzle.[37]

It is important to note that the theory of special creation could accommodate these patterns. Back home, the greatest comparative anatomist in Britain, Richard Owen (father to William Owen, who we met in the introduction) was positing the "unity of plan" between the ancient and modern inhabitants of South America as illustrative of God's "blueprints" for new, successive creations. This explanation allowed naturalists to argue that the fact God used, say, an armadillo blueprint or sloth blueprint, and then varied that form within new exertions of creative power, demonstrated not only divine wisdom and power, but divine efficiency!

Darwin was, for a time, perfectly willing to consider these explanations as both compelling and scientific. Reflecting back on the voyage forty years later, Darwin wrote: "When I was on board the Beagle, I believed in the permanence of Species, but, as far as I can remember, vague doubts occasionally flitted across my mind." Darwin continued: "On my return home in the autumn of 1836, I immediately began to prepare my journal for publication, and then saw, how many facts indicated the common descent of species, so that in July 1837 I opened a note book to record any facts which might bear on the question. But I did not become convinced that species were mutable until, I think, two or three years had elapsed."[38]

The most famous puzzles posed by the voyage arose, of course, from bird specimens Darwin and other crew members collected on the Galapagos Islands. But because Darwin did not actually notice or start pondering these ornithological puzzles until after he got home, we will not discuss them quite yet. Meanwhile, as Fitzroy drew his maps, Darwin still spent most of his time on land collecting specimens, recording his observations, and riding on the coattails of the rapidly expanding infrastructure of European imperialism. In September 1835, four years after leaving England, the *Beagle* crew finally turned the ship west across the Pacific and began the long trip home. Over the next ten months they stopped at islands with names like Otaheite, Aotearoa, Trowunna, and the "island continent" of Australia. Darwin knew these places by the names either mispronounced or conferred by European explorers and colonizers: Tahiti, New Zealand, and Tasmania.

While in South American the *Beagle* had been traveling the routes of what historians have called an "informal Empire" (i.e., places where the British had strong military, commercial, and strategic interests, but in which influence was not exerted via the establishment of colonies or British governmental administration). But in the South Pacific, British imperial control was both formal and explicit. From New Zealand to Australia and on to South Africa, the *Beagle* crew traveled from one British outpost to another. Indeed, Fitzroy had been commanded by the Admiralty to exact payment of a fine from the queen of the Society Islands, Pomare II, to the British government. By what right did men from a small island in the northern Atlantic arrive on faraway shores and declare the land property of the British Crown; carve that land into homesteads; and call upon settlers from England, Scotland, Ireland, and Wales to make it their home? Or, in New South Wales, use that land as a penal colony?

British legislators, diplomats, missionaries, military officers, and settlers drew upon a range of justifications and assumptions in reply. Expanding material wealth was, of course, a primary drive of imperialist thinking and action. The Industrial Revolution needed raw materials to run. But defenders of imperial expansion also saw colonization in moral terms (based on their own standards, of course). After Parliament passed the Slavery Abolition Act in 1833, many British believed that Britain could—and indeed, must—teach the rest of the world how to be moral and fight sin. Britain, they assumed, had been given the special task, by God, of remaking and redeeming the world. Christianity, civilization, and capitalism would form the three-pronged, tightly intertwined means of that process. After abolition, colonialism's defenders argued that Britain was morally obligated to expand its empire in order to guide "heathens" toward both civilization and salvation. And while the Crown may have lost New England in 1776, Britain still had control over much of the West Indies and Canada, and formally claimed more land throughout the nineteenth century (the Cape of Good Hope in South Africa in 1815, British West Africa in 1821, New Zealand in 1840, Hong Kong in 1841, India in 1858, and large swaths of the African continent during the Berlin Conference of 1884–1885). Meanwhile, expanding imperial control and settler-colonialism provided a pressure valve that dampened the threat of political revolution at home. (As we will see, this was particularly the case after a series of failed European revolutions in 1848.)

Imperial and colonialist movement into faraway lands relied, of course, on the assumptions that first, Europeans could group human beings within natural or cultural groups, and second, that they could place those groups within a hierarchy, with European civilization and people at the pinnacle. Some British observers did wonder what kind of safety colonialism posed for the people already living on these lands. Some even questioned whether British civilization was so superior to the societies and civilizations most colonists were intent on replacing. Indeed, the *Beagle*'s very own Auguste Earle, who served as the ship's artist during the first part of the voyage, published a damning portrait of the effects of British settlers and missionaries on Indigenous life in his 1832 book *A Narrative of Nine Months' Residence in New Zealand*.

Both Fitzroy and Darwin were very interested in debates taking place over the influence of British missionaries. Ten years before they arrived in the South Pacific, a Russian explorer named Otto von Kotzebue had written a scathing critique of English missionaries' influence in the South Seas. He accused the missionaries of destroying the islanders' way of life and turning them into gloomy Christians. When the *Beagle* finally landed in Tahiti, and Darwin met "the finest men I ever beheld" (referring to the Tahitians), neither he nor Fitzroy found evidence that Kotzebue's claims were true. Darwin reported home that the Tahitians seemed happy and were advancing in civilization under the missionaries' guidance. He was particularly impressed by the fact that on a trek inland, his guides said a prayer before going to sleep. Furthermore, he declared the three missionaries at the station at Matavai "sensible agreeable gentlemen."[39]

Both Darwin and Fitzroy continued to side with missionaries as they traveled to the mission stations on the Bay of Islands in New Zealand. They compared their impressions with Auguste Earle's harsh criticisms. "We are quite indignant with Earle's book," Darwin reported to Caroline. Proceeding to Australia, he was impressed with how similar Sydney was to home, and saw no moral problem with the fact the land was being turned into a mirror-image of English society. While he had had little good to say about the Maori of New Zealand, he was charmed by Indigenous Australians. But he was very clearly impressed *most* by his fellow Englishmen. Visiting the settlement of Hobart, in Van Diemen's Land (Tasmania), he wrote home that if he ever emigrated it would be there, and then gushed: "It is necessary to leave England, & see distant Colonies, of various nations, to know what wonderful people the English are."[40]

Travelling on to Cape Town, South Africa, Fitzroy and Darwin heard rumors of continued criticisms of local missionaries, criticisms with which both were clearly unsympathetic. Indeed, as they sailed away from the colony toward home, they composed a joint letter about their positive impressions of the missionaries in New Zealand and Tahiti and sent it to the *South African Christian Recorder*. This defense of Anglican missionaries was Darwin's first publication. As historian Janet Browne notes, the letter shows that both men subscribed "to a particular view of human nature, a view based on the concepts of racial unity and a capacity for progress through education and the alleviation of external circumstances."[41]

In retrospect, of course, Europeans' righteous confidence that material and moral progress inevitably accompanied the spread of civilization and private property (imperialists assumed these things went together, and that the only civilized society was Christian), justified much that caused, rather than ameliorated, suffering. As we will see, even Darwin sometimes noticed that fact. But in general, he maintained his support for missions even after, in his forties, he gave up belief in Christianity itself. And he would always see religion and belief in God as having played a crucial role in humanity's moral progress.

At first glance debates over British economic, missionary, and settler expansion may seem to have nothing to do with natural history. Of course, Darwin, and any other British naturalist or collector, traveled along the infrastructure being built through British imperialist ambitions. But the connections went even deeper. Throughout the South Pacific, the interactions between Indigenous and European creatures raised profound puzzles for a devout adherent of the theory of special creation: in island after island, European pigs, dogs, and rats were exterminating the fauna that originally, or "aboriginally" (from the beginning), lived there. In other words, organisms "specially created" for completely different "conditions of existence" (as Cuvier would say) were somehow moving in and outcompeting organisms that an all-wise, all-powerful, and all-good God had supposedly designed for these far-off lands.

These puzzles were heightened by anecdotes and rumors that the pattern of competition and extermination was also occurring within populations of God's most favored species: *Homo sapiens*. First rumors about and then data on mortality rates showed that non-native children were surviving at higher rates than indigenous children. Within a few years (1837) the physician Thomas Hodgkin would establish the Aborigines Protection Society in an attempt to halt whatever it was that, by the 1830s, was clearly decimating Indigenous populations from the United States to Australia. (A devout evangelical certain that it was a Christian duty to convert heathens, Hodgkin did not argue that British missions, physicians, or settlers should just stay home.) But what precisely was happening, and why? And why would God permit these differences in mortality rates? What purpose could they possibly serve?[42]

Sometimes the enormous impact on the mortality rates of Indigenous peoples clearly arose from active intention, as in the extermination campaigns of General Rosas on the Argentine pampas, noted earlier. Elsewhere, as Darwin wrote in his account of the voyage, "besides the several evident causes of destruction" (he listed the introduction of alcohol, European diseases, and the gradual extinction of wild animals) "there appears to be some more mysterious agency generally at work." Writing of Australia, where all recognized "the number of aborigines is rapidly decreasing," Darwin added: "Wherever the European has trod, death seems to pursue the aboriginal. We may look to the wide extent of the Americas, Polynesia, the Cape of Good Hope, and Australia, and we find the same result." Clearly, some did not believe the reports of differential mortality rates because Darwin felt obliged to provide a citation for his claims. He called upon the authority of Rev J. Williams's *Narrative of Missionary Enterprise* for the fact that disease stalks the first intercourse between natives and Europeans, and in a long footnote, he countered skeptics by pointing to the prevalence of similar beliefs among natives themselves.[43]

As with his travels in Brazil, children figured large in Darwin's memories of what was happening in the South Pacific. He wrote of how "It was melancholy at New Zealand to hear the fine energetic natives saying, that they knew the land was doomed to pass from their children. Every one has heard of the inexplicable reduction of the population in the beautiful and healthy island of Tahiti since the date of Captain Cook's voyages." Observing how the original inhabitants of Australia seemed to welcome the advance of the white man because they could borrow dogs from the farm houses and barter for milk from the cows, Darwin thought them "blinded by these trifling advantages." Meanwhile the white man "seems predestined to inherit the country of his children."[44]

Having been convinced of the patterns of loss, Darwin clearly struggled with how to explain those patterns. In his account of how, in a period of thirty years, the natives of Van Diemen's Land (Tasmania)—an island nearly as large as Ireland—were rounded up and moved to Flinders Island, where their population declined precipitously, Darwin explicitly blamed his own countrymen. When noting that "this most cruel step" had become unavoidable on account of the "fearful succession of robberies, burning, and murders, committed by the blacks," he held "no doubt, that this train of evil and its consequences, originated in the infamous conduct of some of our country." But at other times he declared the cause of differential mortality rates between different "varieties of man" a mystery. That mystery stayed with him as the *Beagle* sailed on. Meanwhile, Darwin noticed that it wasn't just "the white man alone that thus acts the destroyer." He recorded that the Malay Polynesians drove before him "the dark-coloured native" in the East Indian Archipelago. Thus, as he wrote in his account of his island travels: "The varieties of man seem to act on each other in the same way as different species of animals—the stronger always extirpating the weaker."[45]

When writing of the "stronger always extirpating the weaker," as he did in the 1830s, Darwin did not either say or imply that he approved of what he described. After all, he clearly did not believe there was anything inherently good about being stronger. "The meek shall inherit the earth" was a fundamental premise of Christianity, no matter how much that premise seemed to be in conflict with the fallen state of creation. And creation had clearly fallen quite far. "It is melancholy," Darwin wrote, "to trace how the Indians have given way before the Spanish invaders." He was appalled by General Rosas's campaign to subdue, and if not subdue exterminate, the Indigenous people of the Argentine frontier. "Everyone here is fully convinced," Darwin mused in his *Journal of Researches*, "that this is the most just war, because it is against barbarians. Who would believe in this age that such atrocities could be committed in a Christian civilized country?"[46]

During the voyage, Darwin was plagued by questions regarding the patterns he observed rather than answers. How to explain why human beings are so inhumane to each other? Why did some groups die and others live? What purpose did all this suffering serve? One might blame all on humanity's sinful nature and its inevitable tendency to corrupt anything humans touched. But for a young man who had been charmed by William Paley's argument that "it is a happy world after all," such fatalism was not appealing. As a naturalist steeped in the theory of special creation, Darwin had been trained to imagine purpose and meaning in every affliction. He had been trained to believe in moral and material progress. Both commitments implied that God must have some plan, some purpose, in letting fellow human beings disappear. (Unlike some of his contemporaries, both evolutionist and otherwise, Darwin did not doubt Indigenous peoples' humanity.) But what, he wondered, could that purpose possibly be?

THE WHOLE FABRIC TOTTERS AND FALLS

As the *Beagle* finally headed home in 1836, Charles Darwin pondered the puzzles he had encountered on the voyage, from fossilized armadillos to the deadly interactions among his own kind. He had seen native plants and animals (supposedly "perfectly created" to fit their environments) supplanted by European species. How to explain these phenomena? Was this all part of God's plan? And why would God have done things this way? In a culture where both ministers and naturalists explained suffering via the benevolent plan of a close personal God, the tremendous suffering of enslaved men, women, and children; the "beastly" behavior of slave-owners; and the extinction of entire peoples in the wake of European settlement posed profound problems.

Meanwhile, Darwin pondered the patterns he had observed amid heated debates over the criteria by which scientific theories must be judged. By what criteria must naturalists, for example, assess different theories of the origin of species as compelling explanations of the puzzles and patterns they observed while studying animals and plants? The most influential philosopher of science of the age, William Whewell, argued an explanation was more likely to offer the "true cause" of phenomena if it resulted in a "consilience of inductions" (i.e., the unification of a number of different facts or patterns under a broad, overarching theory). To Whewell, notes historian Piers Hale, the unifying power of scientific explanations contained a message about God, for "more and more phenomena might be brought under ever greater generalization, ultimately pointing to the existence of one ultimate First Cause."[47] According to Whewell, the most scientific theory must also be the most *consilient* theory, in that it brought the widest range of facts within the purview of its explanatory power. As he traveled, Darwin amassed observations and experiences that he would eventually test against Whewell's criteria by asking whether the theory of special creation or that of evolution offered the most consilient explanation of the world. (The alternative criteria, namely that there must be experimental demonstration to claim the discovery of a "true cause," was not, as we will see, very useful to Darwin.)

Darwin's first months at home were a whirlwind of activity. He had to write up his part of the official account of the voyage and disperse his specimens to experts who would tell him whether he had brought home any new species. He also wished to write a book on the geology of the voyage. By the time he got home, a career as a clergyman had been replaced by ambitions to make a name for himself in science. As we have seen, the two were not mutually exclusive. After all, both his favorite professors at Cambridge, John Henslow and Adam Sedgwick, were reverends in the Anglican Church. But with the model of Charles Lyell, who spent all his time doing geology (with the occasional pious gesture to the first cause) before him, Darwin set aside the idea of becoming a clergyman. (He had not, at this point, given up belief in Christianity.) With his father's blessing, he moved to London, near his brother Erasmus, and got to work supervising the composition and publication of *The Zoology of the Voyage of H.M.S. Beagle* and writing his part of Fitzroy's *Narrative of the Surveying Voyages of His Majesty's ships* Adventure *and* Beagle.

Now, Darwin was not a zoologist. So to compile *The Zoology of the Voyage*, he sent his specimens to experts in London. The fossils went to Richard Owen, the mammals to George Waterhouse, the birds to John Gould, the fish to Leonard Jenyns, and the reptiles and amphibians to Thomas Bell, all of whom composed their respective sections for *The Zoology*. Darwin was fascinated by the conclusions presented to him by these experts. Having collected some birds from the Galapagos, Darwin had thought the specimens represented different *subfamilies* and *families* of birds: perhaps an oriole or blackbird, some finches, and a wren. It was only when the ornithologist John Gould, standing over his specimens at the Zoological Society of London, told him that all his specimens were finches, of thirteen different new species to be exact, that Darwin began noticing the puzzles posed by his Galapagos birds. He also profoundly regretted his inattention to which island each specimen came from!

Faced with Gould's conclusions (and gathering more carefully labeled specimens from his shipmates), Darwin pondered this small group of island birds. As a geologist he knew the islands were relatively recent. Where had this diversity within one group of finches come from? Gould told him they were all from a group distinct to the Galapagos Islands. But why had God created so many new forms on different islands while maintaining the "finch pattern" used on the closest land mass?[48]

The problem posed by the finches would become the most famous of Darwin's puzzles (in part because, amid debates with US creationists in the 1950s, evolutionary biologists cited "Darwin's finches" as proof of natural selection). In fact, the ornithological information that posed the greatest puzzle concerned Gould's conclusions regarding Darwin's specimens of Galapagos mockingbirds. Darwin thought their slight differences in size and color indicated the specimens were varieties of the same species, and presumed their differences to be the result of subtle, slight shifts in the environment (the technical taxonomic point at stake is examined in more detail below). With a more expert ornithological eye, Gould classified Darwin's three specimens (which he did label according to island) as three different *species*, all closely allied with but distinct from the mockingbirds on mainland South America.

The mockingbirds posed a puzzle over geographical space, similar to that posed by fossilized armadillos and sloths over geological time: Why design on the same plan, with slight adjustments in size or color, over geographical space? Why, Darwin wondered, had God not created *one* species of mockingbirds? Why create three species that seemed to split the islands between them? Or, alternatively, why not create a mockingbird for one island and completely different kinds of birds on the others? Meanwhile, as he watched Gould work through his specimens and compared Gould's expert taxonomic conclusions with his own initial guesses regarding his specimens, Darwin was struck by how difficult it was to determine whether a new form was a new variety or a new species. He knew how often taxonomists disagreed with each other's decisions, to the point they were called "species makers" by critics. But watching experts work on his collection brought the difficulties of doing good taxonomy home.

To understand the puzzle confronting Darwin as he watched Gould classify his mockingbirds and finches, one needs to understand something about how taxonomy was done at the time. Taxonomy refers to the naming, description, and classification of animals and plants. Thanks to British natural history's long tradition of seeing the documentation of diversity as pious and important, taxonomists knew a great deal about animals and plants. They knew some species varied a great deal, to the point that it was challenging for taxonomists to know immediately whether a specimen that seemed different from a known form was a new species or variety. The ability to breed had long since provided the best test of whether varieties belonged to the same species, but of course British natural history centered on specimens, which were all dead. So taxonomists had a working rule: if more specimens came in that blurred the boundaries between two presumed species by connecting them along a gradation of variation, then those two species were collapsed into one and the different forms called varieties.

Within the theory of special creation, naturalists assumed God (the first cause) created species independently, while varieties arose from the influence of climate or food on color or size (i.e., via a secondary cause). But within the theory of special creation, varieties never crossed the species boundary; in other words, they never varied so much that a variety could no longer breed with its original form. This was a standard, well-founded, irrefutable Cuverian rule, with no observational or experimental evidence against it. (Dog breeders might produce impressively different dog breeds, but no one had created a new *species* of dog.)

Watching Gould struggling to classify his specimens, Darwin wondered: Why was it so difficult to discern whether a form was a variety or a species, if species are created by God, and varieties arose via secondary causes? Given their very different origins, surely the distinction between varieties and species, to an expert ornithologist, should be easier to discern! These were the moments, bent over specimens after he got home, in which Charles Darwin began doubting whether the theory of special creation provided a truly consilient explanation of the phenomena naturalists studied.

Six months after he returned home from the voyage, Darwin decided that transmutation *might* account for things that seemed to make no sense under the theory of special creation. He opened a small leather-bound notebook, wrote and underlined the word *Zoonomia* (the title of his grandfather's book) at the top of the first page, and began scribbling his thoughts. Transmutation first appeared in the form of a sketch in which ancestral species were at the bottom of a tree-like diagram, and present-day species were at the top. "Heaven knows whether this agrees with Nature," he wrote a few pages later, and then added and underlined one word: *Cuidado* ("be careful" in Spanish).[49]

A range of anxious thoughts may have inspired the one-word warning: perhaps the diagram did *not* agree with nature, and he would be led astray by the fact he was excited about his idea and wished it to be true. (Darwin could be a master at self-criticism: he noticed he tended to remember facts that seemed to agree with his view, and forget those that did not.) Prominent naturalists, from Cuvier to Lyell, had produced strong arguments that species do not change, which meant he would have to systematically refute every one. He would also have to convince his colleagues that the origin of species could be addressed via the methods of science. (Unlike his grandfather, he never tried to explain the origin of life, but left that to the first cause.) All of these concerns were driven by his strong wish, by the end of the voyage, to make a respected name for himself in the sciences. He would have to either follow the rules he learned at Cambridge, or somehow change them.

But the warning to be *cuidado* may also have been inspired by his deep awareness that, by the 1830s, transmutation was associated with radical political ideas. Darwin knew that some of his closest friends and beloved mentors would perceive a defense of transmutation as a betrayal of his own class. For generations, the theory of special creation had been cited to inspire trust in providence as the ultimate source of a highly hierarchical social order. Darwin had plenty of evidence from his time in Edinburgh that alternative explanations of the origin of species would be quickly picked up by those campaigning for radical political and social change. He knew transmutation appealed to those intent on breaking the political power of the Anglican Church, the existing British class system, and barriers to democratic reform. Indeed, soon after he arrived home the unenfranchised had launched a movement to expand the vote (the 1832 Reform Bill had extended the vote, but only to men who owned property) and get the church "out of bed" with the State.[50]

Given how the theory of special creation was supposed to remind individuals of God's benevolent presence amid even the greatest afflictions, any alternative explanation of species also threatened to destroy the entire lesson of the sparrow's fall: God's close personal attention to individual human lives. Paley, Henslow, Sedgwick, Coldstream—all argued that the purposeful design of animals and plants reminded the afflicted that all suffering was sent by a benevolent God for a reason. Even as science's study of secondary causes expanded, Darwin's mentors had all believed that no secondary cause could explain the origin of species. They argued that the purposeful parts of animals and plants provided rational evidence of the truths of Scripture. As a result, anyone who tried to give an alternative explanation of species would be perceived as threatening belief in the God who attended the fall of every sparrow.

The high stakes made Darwin cautious, but he was too fascinated by the potential of transmutation to make sense of the puzzles posed by his own kind and other species to halt. Besides, Christianity (in which Darwin still believed) had adjusted to new revelations about nature before. So the puzzles continued to crowd into his notebooks. The rapid clip of his penmanship shows that he knew he was onto something. More than a hundred and fifty years earlier, John Ray had argued that every part of an animal is fitted to its use: "For neither do any Sort of web-footed Fowls live constantly upon the Land, or fear to enter the Water, nor any Land-Fowl so much as attempt to swim there." Darwin began pondering the existence of web-footed geese that rarely swam, iguana "designed" for desert environments swimming in the sea, and woodpeckers with feet perfected for climbing that lived entirely on the ground.[51]

From his first musings on evolution, Darwin included the mysteries posed by his own kind. The passage about the "fall of the sparrow" in Matthew was as much about human uniqueness as it was about God's close personal attention to individual lives. Christ's words that "Ye are of more value than many sparrows" reminded Christians that God created humans "in His image," above the rest of creation. But within a few short months of opening his notebooks, Darwin had decided it made no sense to talk about organisms as higher or lower. "It is absurd to talk of one animal being higher than another," he jotted. "We consider those, when the intellectual faculties cerebral structure most developed, as highest." A bee, he added, would no doubt consider her own kind the highest of all the animals.[52]

In his second notebook, filled between February and July 1838, Darwin wrote the following in a stream-of-consciousness form:

> *Once grant that species and genus may pass into each other,—grant that one instinct to be acquired . . . & whole fabric totters & falls.— Look abroad, study gradation, study unity of type, study geographical distribution, study relation of fossil with recent. The fabric falls!*[53]

Darwin knew that most naturalists, including Charles Lyell, considered human beings higher creations on account of their rational and moral capacities. But having seen his own kind behave worse than animals, he began imagining that one might posit a gradation in mental apparatus, just as one might see gradation in physical form between apes and *Homo sapiens*. In doing so he lowered humans, who were supposed to be closest to angels, down toward apes and raised apes closer to humans. Reflecting on recent trips to the London Zoo, he imagined a man comparing the orangutang (its "expressive whine," intelligence "as if it understood every word said," and "affection to those it knows") with "savage" man ("roasting his parent, naked, artless, not improving, yet improvable") and demanded: "Let him dare to boast of his proud preeminence."[54]

Intent on consilience, Darwin took *all* behavior into his purview, including that of his own species. How, for example, did human beings convince themselves it was right to enslave their fellow human beings? He made notes on how defenders of slavery wished to make the enslaved man another kind so that they need not consider him equal. "[T]he white Man," Darwin wrote, "who has debased his Nature & violates every best instinctive feeling by making slave of his fellow black, often wished to consider him as other animal." He then turned the argument that designating a man as "another animal" justified different treatment upside down by collapsing the animal-human boundary entirely: "Animals our fellow brethren in pain, disease death & suffering & famine; our slaves in the most laborious work, our companion in our amusements, they may partake, from our origin in one common ancestor we may all be netted together." Of course, this meant the suffering in the world (in marked contrast to Paley's "It is a Happy world after all") was very great indeed.[55]

Within months of returning home, Darwin had convinced himself that transmutation might explain his puzzles better than the theory of special creation. But as a naturalist trained by Henslow and Sedgwick, he knew that unless he could explain—via a purely natural, material mechanism (a secondary cause)—*how* organisms had been modified "so as to acquire that perfection of structure and coadaptation which most justly excites our admiration" (all the purposeful parts described by John Ray, William Paley, and generations of naturalists), there was no point in publishing his speculations.[56] He would need to have a secondary cause that soundly bested Paley's argument that purposeful parts (and therefore species) must have been independently, specially designed (as Sedgwick put it, via the miraculous breathing of "a living spirit" into dead matter). He would need to deal with every one of Cuvier's laws that supposedly precluded the possibility of species change. His argument and his methods would have to be airtight. Ideally, he would also somehow avoid undermining the religious and political commitments of his own class. Given the daunting tasks before him, Darwin was not inclined to rush toward a public confession of his conversion to his grandfather's view of the origin of species.

Chapter Five

Ten Thousand Sharp Wedges

As he scribbled about monkeys and men in his notebooks during his first years home, Darwin was also courting his cousin, a young woman named Emma Wedgwood. They would marry on January 29, 1839, three months after Darwin hit upon the idea of natural selection. Over the course of their long, and by all accounts happy, marriage, Charles and Emma would bury three of their ten children. The first, Mary Eleanor, died on October 16, 1842, at just three weeks old. Annie Darwin lived until the age of ten and died on April 23, 1851. Eighteen-month-old Charles died of scarlet fever on June 28, 1858.

During the latter half of the nineteenth century the high child mortality rate experienced by families of all creeds, colors, and classes in Britain would slowly but steadily begin to go down for some. But few hints of that impending decline existed in the 1840s. Child loss was part of the Victorians' world and the common experience of most parents. By all accounts Charles and Emma supported each other in the wake of these bereavements. "We shall be much less miserable together," Emma wrote to Charles, the day after Annie's death. But like many Victorians, they ultimately found solace in very different explanations of child loss.[1]

A Fearful Mortality among Infants

Darwin returned home on October 2, 1836, to a changed Britain. The 1832 Reform Bill had created a power-sharing agreement between landed wealth and new middle- or upper-class capitalists. Whig manufacturers and wealthy religious dissenters had increased their power, from

Parliament to town councils. An emphasis on both open competition and religious freedom as the foundation of material and moral progress suited the Whig political platform, to a point. Merchants and industrialists favored free trade and open competition in both industry and education, so long as the competitors were male, owned property, and were of European descent. In other words, having obtained more power via legislative reform, the rising middle class called a halt to expanding the franchise further. The new powers-that-be did not seriously consider giving the vote to working men, much less women.

When, unsatisfied with the 1832 Reform Bill, radicals began calling for universal suffrage and a democratic "People's Charter," wealthy Whigs had strong incentives to root the new status quo (and middle-class power) in both God and nature. They needed a vision of nature that would ensure slow, law-bound change rather than miracles (which, according to critics, propped up Tory and Anglican power) or revolution from below (the province of atheist evolutionists insisting that organisms could will extraordinary change from within).

It turns out that the work of Thomas Malthus, who we met in chapter 3 trying to reconcile poverty with God's benevolence, fit Whig ambitions perfectly. Indeed, in 1834, as Darwin wandered around South America, the Whigs had passed a Poor Law Amendment Act that reflected Malthus's vision of the struggle for existence as the basis of progress. The government forbade outdoor charity and instead built notoriously uncomfortable "workhouses" for the poor. Workhouse officers separated husbands and wives (though only at night!) and gave men, women, and children tedious tasks like picking oakum (removing tar from old ropes) on the grounds that miserable working conditions would inspire the poor to, first, find work elsewhere, and second, have fewer children by practicing moral restraint (sexual abstinence).

The workhouse system assumed, of course, that there was work to be had. Plenty of contemporaries noticed that in a boom-and-bust industrial economy this assumption was dead wrong. One critic (who designated the Poor Law Amendment Act the "Malthusian Bill") argued that the act was in fact designed to force the poor to emigrate, or work for lower wages and live on poor-quality food. Other critics pointed out that, with more workers trying to avoid the misery of the workhouses, competition between workers increased and factory owners could pay less. Meanwhile, low wages at home indeed drove emigration to the farthest reaches of a growing British Empire, while defenders of colonialism urged imperial expansion as the best outlet for increasing calls for reforms—or worse, revolution.[2]

By the time Darwin returned home, Thomas Malthus was hated by some and admired by others. "Parson," wrote the reformer William Cobbett, "I have, during my life, detested many men; but never any one so much as you." Charles Dickens famously pilloried the Poor Law Amendment in *Oliver Twist, or the Parish Boy's Progress* (1837). More aggressive critics like Karl Marx and Friedrich Engels would soon join Malthus's old adversary, William Godwin, in arguing that neither nature nor God were responsible for human misery. Rather, they argued that the industrial-capitalist economic system created poverty and suffering. They also argued that the system could and must be changed.[3]

Upon returning home, Charles Darwin lived and worked among ardent proponents of the Poor Law Amendment. Thomas Malthus had been dead two years when Darwin returned home, but it turns out that some of Darwin's brother's best friends kept the old parson's ideas alive. Throughout 1837, Darwin dined at his brother Erasmus's home on Great Marlborough Street in London, where the most influential advocate of Malthus's ideas, Harriet Martineau, also spent her evenings.

Martineau wrote wildly popular books at a time when both God and society tightly circumscribed women's roles. They could be mothers, caregivers, and homemakers. Martineau constantly rebelled against such boundaries. She never married and published book after book, always under her own name. She helped found the disciplines of political economy and sociology. She campaigned for the immediate abolition of slavery as other supposed abolitionists waffled and whined about moving too fast. She defended female education and wrote some of the earliest essays in English about what it was like to become deaf (she lost her hearing at the age of twelve). When Darwin met Martineau, she was a Unitarian, but two decades later she co-wrote the first defense of atheism (she called it necessarianism) published in Britain. Erasmus advised his brother that the only way to make sense of "Miss Martineau" was to remember that "one ought not to look at her as a woman."[4]

Clearly, Martineau had her own ideas, but in the 1830s a large part of her writings argued that Thomas Malthus was right about how nature worked. She was the one who insisted that the "fearful mortality among infants" of the poor had inspired Malthus to write his *Essay on Population* in the first place. And she was intent on spreading his ideas as the best means of both preventing and ameliorating suffering in future. Martineau's presence in Darwin's social circle meant that he was surrounded by individuals debating explanations of why the world is the way it is, whether and how it could be changed, and what must be believed about both God and nature. As he tried to understand the things he had observed during the voyage, he learned of Martineau's campaign for Malthus's understanding of the world, society, and his own kind firsthand.

At first glance, Martineau's support for the work of a devout British clergyman may seem strange. But although they disagreed about God (Malthus was Anglican; Martineau a lapsing Unitarian), Martineau and Malthus agreed on the natural laws that governed the world. In books and articles, Martineau evangelized Malthus's ideas to the British masses while turning Malthus himself into a virtuous savior of humanity. His clear insight into nature's laws, she believed, would save humanity from the self-induced suffering that arose from ignorance. "The desire of his heart," Martineau wrote, "and the aim of his work were that domestic virtues and happiness should be placed within the reach of all, as Nature intended them to be."[5]

Thanks in part to Martineau, Malthusian thought was pervasive within the liberal Whig circles in which Darwin moved. Malthus's explanation of suffering (an explanation, as we have seen, aimed at vindicating God's wisdom, goodness, and power) suited the Whig commitment to industrial-capitalist competition perfectly. Recall that Malthus argued that the existence of great misery and suffering, including disease, resulted in the intellectual progress of man. Suffering (for example, that produced by population inevitably outstripping food supply) inspired the exertion required for the continued progress of the human mind. Malthus thus argued that evil exists in the world "not to create despair, but activity." Indeed, he insisted that every individual had a duty to "use his utmost efforts to remove evil from himself and from as large a circle as he can influence." The more wisely and successfully a man exercised this duty, the more he would improve and exalt his own mind, and in doing so more completely "fulfill the will of his Creator."[6]

Both Malthus and Paley insisted that this system depended on God generally *not* intervening in natural laws. Nature's laws, Malthus insisted, must be uniform. Otherwise, man could neither learn from the consequences of his actions nor prevent suffering in future. He believed, for example, that the poor, instructed in the science of population (i.e., the consequences of certain actions) would turn to moral restraint rather than revolution to solve their problems. Malthus also argued that only through the protection of private property, equality before the law, and civil and political liberty could men believe their efforts would be consistently and fairly rewarded. Like the founders of the new sciences, he was also confident that this emphasis on natural law would not undermine Christianity. Both Scripture and the theory of special creation would maintain God's close personal governance over creation.[7]

When Charles Darwin picked up the sixth edition of Malthus's *Essay on Population* "for amusement" in October 1838, he had, as we have seen, set aside the theory of special creation. Casting about for some alternative means by which species might be created, Darwin found what he needed in Malthus's book. He later recalled that he was already "well prepared to appreciate the struggle for existence that everywhere goes on." As he read Malthus's *Essay*, it "at once struck" him "that under these circumstances favourable variations would tend to be preserved, and unfavourable ones to be destroyed. The results of this would be the formation of a new species." Here, then, Darwin recalled, "I had at last got a theory by which to work."[8]

Darwin was not, of course, the first to imagine that Malthus's struggle for existence might explain natural phenomena. Charles Lyell had used the struggle for existence to explain the extinction of species. He did not write of that struggle as breaking the species boundary and creating new organs or new species. It was Darwin, and Darwin alone (at least in 1838), who turned the struggle for existence into something that might be *creative*. Immersed in the puzzles crowding his notebooks, Darwin imagined that the struggle for existence might also lead to the creation of *new* species.

Darwin may have had a "theory by which to work," but he also, suddenly, had a tremendous amount of additional work to do. He had to determine whether this process, which by 1841 he called natural selection, could both push a population across the species boundary and explain the origin of purposeful parts, including the supposedly unique traits of human beings. In other words, he had to determine whether natural selection provided a consilient secondary cause for what almost all naturalists, including his beloved professors and Charles Lyell, believed required the direct intervention of the first cause, God.

OF SORROW'S LINKS A FIRM CONNECTING HAND

Over the next twenty years, as Darwin tried to determine whether his "theory by which to work" agreed with nature, his views of both religion and God changed. The relationship between his shifting beliefs regarding God and evolution was complicated and certainly not one-directional. Victorian unbelief did not necessarily lead to evolution (much less natural selection), nor did Darwin's theory of evolution necessarily lead to unbelief (as we will see, some of Darwin's most committed defenders after he published his work were evangelical Christians). For Darwin himself, clearly conflict between his prior belief in God's close personal providence and his new, evolutionary vision of nature existed, but what precisely the conflict was about was neither static nor straightforward.[9]

In 1876, long after publishing *On the Origin of Species*, Darwin reflected on how his views of God and religion had changed over his lifetime. He wrote that during the two years after the voyage, "I was led to think much about religion." He was slowly abandoning Anglican doctrines, but as we have seen, plenty of other versions of Christianity existed, including Unitarianism. (By his own account, Darwin did not abandon Christianity for more than a decade *after* he adopted evolution.) He also wrote that he "did not think much about the existence of a personal God until a considerably later period of my life." So he thought much about some things concerning religion and not much about other things during his first years home.[10]

Darwin had good reasons to "think much about religion" between 1836 and 1838, even apart from pondering how evolution might fit with Anglican or Unitarian assumptions about the world. For despite his fears over the "terrible loss of time" and "the expense & anxiety of children" that marriage entailed, he was courting Emma Wedgwood, by all accounts a devout Christian. Charles and Emma both came from highly eclectic religious backgrounds. As first cousins they shared many of the same family traditions, including an easygoing mix of Anglican respectability and a more radical tinge of Unitarian dissent. But during their courtship they both clearly sensed they might be on very different paths amid the various kinds of religious beliefs in the Darwin-Wedgwood households.[11]

During his two years thinking much about religion, for example, Darwin had certainly not solidified his faith in the Thirty-Nine Articles of the Anglican Church. At some point he must have confessed some skepticism to his father, Robert Darwin, who warned him not to share his doubts with his future wife. Dr. Darwin explained that he "had known extreme misery thus caused with married persons. Things went on pretty well until the wife or husband became out of health, and then some women suffered miserably by doubting about the salvation of their husbands, thus making them likewise to suffer." Both father and son assumed that unbelief was rare among women (Dr. Darwin attested that he had only known three women skeptics). As historian Evelleen Richards notes, religion was a highly gendered affair within Victorian middle- and upper-class homes, and the Darwin and Wedgwood families did not depart from this pattern. Doubt might be tolerated in men; it was not only impious but unnatural in women. (Recall that Miss Martineau was not, according to Erasmus, to be seen as a woman.)[12]

Despite both his father's advice and his own assumptions about female religiosity, Charles refused to be secretive on such an important matter. We do not know what doubts he confessed, but we have Emma's thoughtful reply, which she wrote shortly after their engagement: "I thank you from my heart for your openness with me. My reason tells me that honest & conscientious doubts cannot be a sin, but I feel it would be a painful void between us." She asked him to read her favorite part of the New Testament: Christ's farewell to his disciples.[13]

Emma believed in Christ's vicarious atonement for human sin. She believed in heaven and in faith in Christ as the path of salvation. But unlike William Paley or Adam Sedgwick, she did not appeal to woodpecker beaks or earwig wings to demonstrate God's care. She did not think matters of faith were demonstrable by reason, in the same way Darwin might weigh different explanations of the origin of species (or whether to marry!). After they were married, she wrote what she could not say in person, namely that she wondered whether "the line of your pursuits may have led you to view chiefly the difficulties on one side, & that you have not had time to consider & study the chain of difficulties on the other, but I believe you do not consider your opinion as formed." She tried to remind him that the truths of Christianity might not be weighable by the same criteria as matters of natural history: "May not the habit in scientific pursuits of believing nothing till it is proved," she wrote, "influence your mind too much in other things which cannot be proved in the same way, & which if true are likely to be above our comprehension." A great deal was at stake, in Emma's view. She concluded her letter: "Don't think that it is not my affair & that it does not much signify to me. Every thing that concerns you concerns me & I should be most unhappy if I thought we did not belong to each other forever."[14]

Ultimately, Emma and Charles reached a respectful truce on the matter. Henrietta Darwin later described her mother as "sincerely religious" and someone who found her only relief from anxiety in prayer. When they were young, "she went regularly to church and took the Sacrament. She read the Bible with us and taught us a simple Unitarian Creed, though we were baptized and confirmed in the Church of England." This was a common enough strategy among manufacturing elites like the Wedgwoods. As historian Janet Browne notes, the family "willingly embraced the conventional outlook of the Church of England" as they turned themselves into landed Whig gentry. "Religion," notes Browne, "was as much a question of class and behaviour as doctrine."[15]

Emma did not like Anglican doctrine, however. By the time she taught her children "a simple Unitarian creed," Unitarianism had become less radical than that of its founder, the chemist Joseph Priestley, who had denied the divinity of Christ. Yet on balance, Unitarians were extremely tolerant of both new scientific ideas and emphasizing God's governance via natural laws. They emphasized God's benevolence (over "inscrutable justice"), which in turn inspired faith in God-given reason to discern causation in nature rightly. Furthermore, their suspicion of formulated creeds and codified doctrine made it difficult to establish some kind of rule that would limit natural explanations to certain realms. Meanwhile, Unitarians upheld Anglican Christianity's traditionally progressive vision of history in which both spiritual *and* material progress demonstrated God's benevolent dispensation toward humanity. To believe, as Charles increasingly did, that "the more we know of the fixed laws of nature the more incredible do miracles become" was a perfectly fine stance for a liberal Protestant to take, so long as certain exceptions remained: a divinely endowed soul through which God conferred rational and moral capacity (and responsibility) upon humanity, and the promise of salvation. The trouble was, of course, that Charles had abandoned belief in the tight boundary between humans and animals upon which those exceptions were generally supposed to depend.[16]

Despite their growing differences, Emma arranged their home so that Darwin's working hours were disturbed as little as possible as the Darwin babies were born and grew into rambunctious children. By 1842 Charles felt confident enough to sketch a thirty-five-page outline of his theory. That fall, the two eldest children, William and Annie, were just one and two years old, respectively. In September Emma gave birth to a girl, Mary Eleanor, shortly after the family moved from London to "Down House" in the country. The little infant lived less than a month. Four days after her baby died, Emma replied to a "sweet, feeling note" from her sister-in-law Fanny Wedgwood. "Our sorrow is nothing to what it would have been," Emma wrote, "if she had lived longer and suffered more." "With our two other dear little things," Emma concluded, "you need not fear that our sorrow will last long, though it will be long indeed before either of us forget that poor little face." No one recorded what Mary Eleanor died of (the cause of death was not routinely included in death records until much later in the century).[17]

The experience of losing an infant was common enough, but no less painful or easy to explain for that. We know from Emma's letter to Charles regarding his "honest doubts" that their explanations of child loss were already diverging. Applied to the loss of a small infant, the differences between Charles and Emma's trajectories of faith eventually became quite deep. Some years later, Emma wrote Charles a letter in which she confessed that she found the only relief amid affliction "is to take it as from God's hand, & to try to believe that all suffering & illness is meant to help us to exalt our minds & to look forward with hope to a future state." She then reminded him that "it is feeling & not reasoning that drives one to prayer," and was not ashamed to imagine that feeling, sentiment, and a desire for comfort might be the source of belief. That one could explain what drove humans to prayer did not, in her mind, therefore justify doubt. (Darwin wrote "God Bless you. C.D. June 1861" on the letter before filing it away.) Darwin later claimed that he had not given up Christianity in the 1840s, but he clearly wished to reason about such things, rather than be guided by feeling. And that meant that, as the years passed, and Emma explained the most terrible afflictions as "from God's hand," Charles tried to find an explanation solely in nature's laws, whether those laws were designed by God or not.[18]

There was one explanation that neither Charles nor Emma had to consider—namely, that their baby died because of poverty. In 1845 the poet Caroline Norton cited child loss as evidence that, although access to pleasures divided the rich and the poor, "the Divine command hath made of Sorrow's links a firm connecting hand." Poor mothers grieved "the shadow of a child" on her knee with the same sorrow with which duchesses "pray'd through many a midnight watch in vain."[19] But by the time Norton penned her poem, some were noticing that such sorrows did *not* seem equally distributed at all. Between 1832 and 1842 a series of government reports exposed the chilling conditions in which children as young as four worked in the Irish, Scottish, Welsh, and English mines and factories driving the Industrial Revolution. Poet Elizabeth Barrett Browning turned the reports into an imaginary account of the lesson these children learned about God from the machines and men around them:

Do not mock us; grief has made us unbelieving—
We look up for God, but tears have made us blind.
Do ye hear the children weeping and disproving,
O my brothers, what ye preach?[20]

We have seen how by 1840, the Scottish physician Andrew Combe used evidence of different infant mortality rates between the rich and the poor to argue that better "infant management" (i.e., infant care) by both parents and physicians was possible. Some called for a revolution in the management of society itself.

The same month little Mary Eleanor Darwin died, a radical socialist named George Jacob Holyoake and his wife Helen also experienced the loss of a beloved daughter. At the time, Holyoake was serving a six-month prison sentence in Gloucester for "uttering certain blasphemous words against God." During a lecture for the Cheltenham Mechanics Institute criticizing the Poor Law Amendments, a member of the audience had asked Holyoake where man's duty to God fit into his proposed reforms. Holyoake replied by pointing out that amid tremendous national debt, the national church and religious institutions cost the nation about twenty million annually. "Worship being thus expensive," he then said, "I appeal to your heads and your pockets whether we are not too poor to have a God?"[21]

When Holyoake was imprisoned, the Anti-Persecution Union sent ten shillings a week to Helen to feed her and their two daughters, but it was not enough. Holyoake was certain that malnutrition and poverty had killed his daughter. Madeline "had died the death of the poor," he later wrote. "She had perished among the people who knew neither hope nor comfort." In subsequent months he pledged his life to those "whose sad destiny one so dear to me is linked." "Though in the death of poverty there is nothing remarkable," he wrote, "though hundreds of children are daily killed off in the same way, yet parents unused to this form of calamity find in it, the first time, a bitterness which can never be told."[22]

What did any of this have to do with how men explained the origin of species? The answer turns out to be "Plenty." After Madeline's death, while he was still in prison, Holyoake received a package from a man named Reverend Andrew Sayer. Inside he found a copy of Paley's *Natural Theology: or, Evidences of the Existence and Attributes of the Deity*. Holyoake spent the rest of his imprisonment pouring his anger and grief into a pamphlet entitled *Paley Refuted in His Own Words*. His book came off the printing presses of a radical publisher named Hetherington in 1843. "Paley professes to show that experience" of, say, the fact watches are made by watchmakers demonstrated the rational certainty of God's existence. But one must extend the analogy: Who, then, designed this Grand Watchmaker, God? As for Paley's determinations of God's attributes, Holyoake was merciless: "The numbers and the poignancy of human sufferings are bitter truths, which no sophistry can hide. The weight of evil in the world must crush the system of the natural theologian." Paley, he argued, "meddles with a subject which exceeds human comprehension." Nothing could be known about the existence of a deity other than that "nothing can be known." At the end of the decade Holyoake coined a word for the stance that man must make his way in the world without God: secularism.[23]

The Darwin family also lost children, but they were completely insulated from having to imagine, as Holyoake did, that their bereavement was the result of an unjust social system. Indeed, the Darwin and Wedgwood fortunes were built on the industrial-capitalist economic system Holyoake and company were intent on destroying. Having amassed great capital, the Darwins had access to plenty of food and the best physicians (although even the best could still do very little). By the mid-1840s, those physicians could administer the extraordinary new discoveries of the nonaddictive anesthetics chloroform and ether. Darwin himself administered chloroform during at least one of Emma's pregnancies; it was "the grandest & most blessed of discoveries," he crowed.[24] They tended to see law-bound progress all around them, even if, as Darwin firmly believed, that progress arose via a Malthusian struggle for existence. But Darwin had no interest in supporting radicals, revolutionaries, or atheists like Holyoake by imagining that child mortality arose from an unjust social order, rather than nature's ways.

An Answer to a Certain Extent Satisfactory

In 1844, Darwin confessed to his friend, the botanist Joseph Dalton Hooker, that he had hit upon a secondary cause for the origin of species. (He was careful to point out that his means of change were wholly different than Lamarck's.) Masking his confidence behind humility, he called it a "very presumptuous work & which I know no one individual who w^d not say a very foolish one." He insisted he had collected "blindly every sort of fact," and that "At last gleams of light have come, & I am almost convinced (quite contrary to opinion I started with) that species are not (it is like confessing a murder) immutable." Furthermore, he thought he had found out "(here's presumption!) the simple way by which species become exquisitely adapted to various ends."[25]

Hooker had his own doubts about the ability of the theory of special creation to explain botanical patterns, and his reply was encouraging. Seven months after his "confession," Darwin wrote out a 231-page essay outlining his theory. He attached a letter to Emma, requesting that in the event of his death she ask one of various friends to publish it ("Lyell would be the best if he would undertake it," he noted). As he wrote this "1844 Essay," the footsteps of small children pervaded the Darwin's country home, Down House. William was five. Annie was three. Henrietta was one. And George was on the way. He still thought he had plenty of time.[26]

Then, in October 1844, three months after Darwin finished his essay, a book called *Vestiges of the Natural History of Creation* appeared on the shelves of British booksellers. The book explained the patterns in the fossil record, comparative anatomy, and embryology (all the observations carefully amassed to demonstrate God's close personal control over creation) as the result of a purely mechanistic, naturalistic, and evolutionary "Law of Development." No author's name accompanied the title page.

The man who wrote *Vestiges*, Robert Chambers, did not think putting his name to *Vestiges* worth the potential danger to his reputation or business. Theories of evolution were still associated with both political sedition and atheism, and Chambers's publishing company depended on the loyalty and trust of Scottish evangelicals, who bought many of their textbooks. He also had no formal training in either natural history or natural philosophy, and there was no point in drawing attention to his lack of qualifications to write about either theology or natural history. His wife Anne, who carefully copied out the entire manuscript of *Vestiges* and all subsequent revisions so her husband's handwriting would not be recognized, agreed that the strategy of anonymity was best.

When, years later, Chambers was asked by his brother-in-law why he never acknowledged having written *Vestiges*, Robert replied by pointing to his house, in which he had eleven children, and then slowly added, "I have eleven reasons."[27]

There had almost been fifteen. But four children born to Robert and Anne had died when very young. In 1833 an unnamed child died a few hours after it was born. In March 1842, their two-and-a-half-year-old daughter Margaret died of scarlet fever. Six months later another daughter, just three weeks old, died. The baby must have been a twin, for in 1843 a boy named William died before reaching one.[28]

We know from the pages of the journal Chambers edited that such losses were devastating. In 1832 he printed an essay entitled "Children" that included the line, "A child is, in one sense, a dangerous possession: It is apt to warp itself into the vitals of our very soul; so that, when God rends it away, the whole mental fabric is shattered." It was during the two years after Margaret's death that Chambers poured out his own view of "divine governance" in *Vestiges*. After a whirlwind tour of the history of the earth, from the formation of planets to the appearance—and occasional disappearance—of mammals in the fossil record, Chambers declared that the widely accepted view that God created species through independent exertions did not "comport with what we have seen of the gradual advance of species, from the humblest to the highest" as displayed in the geological record. Chambers then insisted that there must be some other mode by which the "Divine Author" created the organic world. "What is to hinder," he asked, "our supposing that the organic creation is also a result of natural laws, which are in like manner an expression of his will?"[29]

Chambers argued that the "whole train of animated beings" was, since the dawn of life, created via spontaneous generation (he called it a "chemico-electric operation") and subsequent "advances of the principle of development." Advances in form arose due to the force of certain external conditions, namely air and light, upon the developing embryo. Well aware of his contemporaries' fear that explaining the origin of species through a secondary cause would threaten the "dignity of man," Chambers urged there was nothing to be ashamed of in animal ancestry: "It hath pleased Providence to arrange that one species should give birth to another, until the second highest gave birth to man, who is the very highest: be it so, it is our part to admire and to submit." He also insisted that creation via a developmental law, rather than independent exertions, led to a more exalted view of the Divine, on the grounds "it is the narrowest of all views of the Deity, and characteristic of a humble class of intellect, to suppose him acting constantly in particular ways for particular occasions."[30]

Most importantly, Chambers believed that his "Law of Development" explained much that seemed wrong or nonsensical in the world. Take the useless mammae of the male human being that had puzzled John Ray and William Paley. Under the theory of special creation, such organs "could be regarded as in no other light than as blemishes or blunders—the thing of all others most irreconcilable with the idea of Almighty Perfection which a general view of nature so irresistibly conveys." Create species via natural law, and such "abortive parts" became "harmless peculiarities of development, and interesting evidences of the manner in which the Divine Author has been pleased to work." (Chambers had personal experience with "abortive parts": he had been born with six fingers on each hand and six toes on each foot. A bungled amputation to remove the extra digits—before the days of effective anesthetics, mind—left "delicate protuberances" on his feet, which his brother described as "calculated to be a torment for life" and which rendered him permanently lame. One can imagine a young Robert Chambers puzzling over why a benevolent God had allowed such evil to descend on an innocent embryo, and then searching from a young age for a satisfactory answer.)[31]

Chambers extended the ability of evolution to explain things to the most difficult and tragic elements of human experience. What bearing, he asked, did this hypothesis have on the question, asked by the sage "in every age . . . How . . . should a Being so transcendently kind, have allowed of so large an admixture of evil in the condition of his creatures?" Chambers answered with confidence: "Do we not at length find an answer to a certain extent satisfactory, in the view which has now been given of the constitution of nature? We there see the Deity operating in the most august of his works by fixed laws, an arrangement which, it is clear, only admits of the main and primary results being good, but disregards exceptions." Acting independently "each according to its separate commission," such laws can only have effects that are *generally* beneficial, since often one law will interfere with another, and evil will result. As illustration, Chambers provided the example of a boy who, engaged "in the course of the lively sports proper to his age," suffered a fall that crippled him for life. "Two things have been concerned in the case," Chambers explained, the love of exercise and the law of gravitation. Both things, designed by God, were in the main good. "But when it chances," he wrote, "that the playful boy loses his hold (we shall say) of the branch of a tree, and has no solid support immediately below, the law of gravitation unrelentingly pulls him to the ground, and thus he is hurt. . . . The evil is, therefore, only a casual exception from something in the main good."[32]

As we have seen, Chambers's explanation of the falling boy as the unfortunate result of benevolent laws had appeared very early in the history of the new sciences. The seventeenth-century chemist Robert Boyle had written that he thought it "perhaps unreasonable" to expect God to intervene in natural law to save an individual (to suspend, for example, the law of gravity when someone fell over a cliff). But Boyle had called upon both Scripture and God's purposeful design of animals and plants to regain what one might potentially lose in such an explanation: God's care for individuals and belief in a higher system of rewards and retributions. Paley, too, had appealed to the uniformity of nature's laws as the most comprehensive solution to the problem of evil. But he, too, insisted that the purposeful design of animals and plants proved that, in a world primarily governed by divinely designed natural laws, God still attended the fall of every sparrow. Chambers, by contrast, drove natural law into the origin of species and, given the tight ties that had been drawn between the theory of special creation and special providence, threatened the consensus regarding what one must believe about God and why.[33]

Chambers described the benefits of relying solely on secondary causes as improving mankind's ability to both explain and eventually ameliorate suffering. "The Great Ruler of Nature," Chambers wrote, "has established laws for the operation of inanimate matter, which are quite unswerving, so that when we know them, we have only to act in a certain way with respect to them, in order to obtain all the benefits and avoid all the evils connected with them." In other words, Chambers acknowledged that great suffering existed, but argued that the first cause had benevolently provided the means of escape from evil and suffering by making nature's laws uniform. Once, for example, man saw the human constitution as merely a complicated but regular process in electro-chemistry, the path toward elimination of disease—"so prolific a cause of suffering to man"—became clear: to learn nature's laws, and to obey them. Chambers had to admit, of course, that mortality tables still showed a "prodigious mortality among the young," but he was confident that "to remedy this evil there is the sagacity of the human mind, and the sense to adopt any reformed plans which may be shown to be necessary." At first glance this was standard fare within even orthodox Anglican attitudes toward science, but Chambers departed from his predecessors in arguing that divine benevolence must be witnessed *not* in Scripture's promise of an afterlife, but solely in the extraordinary possibilities of future progress as men learned the natural laws that governed the world. (The book's conclusion expressed hopeful uncertainty, but uncertainty nonetheless, regarding the existence of heaven.)[34]

But was Chambers's deity the God of Christianity? And was his theory scientific? Reverend Adam Sedgwick, Darwin's beloved geology professor, certainly did not think either question could be answered in the affirmative. The book was so unscientific, Sedgwick wrote, that he originally thought a woman had written it! As for the author's theology, Sedgwick called his conclusions "seductions," "the serpent coils of a false philosophy," and "rank, unbending, and degrading materialism." *Vestiges* would teach England's women, he wrote, that their Bible is a fable when it teaches men are made in the image of God. (Chambers explicitly imagined humans arising from apes rather than Adam and Eve.) Disturbed by the book's popularity, Sedgwick insisted that the public must be boldly told: nature "gives us a personal and superintending God, who careth for his creatures."[35]

The geologist Hugh Miller backed Sedgwick up while making the stakes clear: *Vestiges*, Miller wrote, belonged to a "school of infidelity" that set wholly aside "the belief in a special Providence, who watches over and orders all things, and without whose permission there falleth not even a 'sparrow to the ground.'" Such infidels replaced belief in a providential God with "belief in the indiscriminating operation of natural laws; as if, with the broad fact before them that every man can work out his will merely by knowing and directing these laws, the God by whom they were instituted should lack either the power or the wisdom to make them the pliant ministers of *his*."[36]

Sedgwick and Miller both hoped to ridicule the book into oblivion on both scientific and theological grounds, but instead *Vestiges* became the nineteenth century's first bestseller. Fourteen editions were published in Britain alone. Prince Albert read it to Queen Victoria. Abraham Lincoln read it. Charles Darwin read it. (He wrote to Hooker, no doubt with some relief, that "His geology strikes me as bad, and his zoology far worse.") More than eighty reviewers summarized and responded to the book in magazines, journals, and newspapers.[37]

Very few readers took the mechanism Chambers proposed for the evolution of species seriously. (Meanwhile, a young man named Alfred Russel Wallace left for the Amazon in an attempt to find the right one.) But reformers loved its message of the duty of human action in the face of suffering. Radicals loved the book's emphasis on the role of the environment in moving an organism forward or backward. (*Vestiges* argued that monstrosities and defects were "the result of nothing more than a failure of the power of development in the system of the mother, occasioned by weak health or misery." Environmental conditions were key, and led to either degeneration or progress: "Give good conditions, it advances," *Vestiges* claimed, "bad ones, it recedes.") A review in the radical medical journal *The London Lancet* called the book's speculations "a breath of fresh air to the workmen in a crowded factory." If the book made men "wander" in the realm of speculation too much, the reviewer wrote, "they had much better wander than sit still in the shadow of apathy which has so long *leadened* (to coin a heavy word) the scientific public."[38]

For those rebelling against orthodox Christianity, *Vestiges* offered an alternative to the theory of special creation and, in doing so, undermined the Anglican doctrines and power to which that theory had been tied for generations. The book offered an alternative explanation of suffering *and* seemed to heighten the possibility of improving the human condition. But while dissenters, including some Unitarians, proved to be some of *Vestiges*'s greatest allies, even they argued over whether the book's emphasis on natural law put God too far off and undermined belief in special providence.[39]

In a lengthy reply to his critics entitled *Explanations*, Chambers tried to make the stakes of choosing between the theory of special creation and the "law of development" clear: "Daily health and comfort, life itself," he wrote, "are sacrificed through the want of this knowledge" of the laws by which the first cause governed the world. We did not become sensible of this truth in hours of cheerful, active, and prosperous existence, he wrote. Rather, we must seek convictions on this subject in more painful hours, during which suffering resulted from a misapplication of human reason to the requirements of both health and happiness. We must look for the truth "beside the death-beds of amiable children, destroyed through ignorance of the laws of health, and hung over by parents who feel that life is nothing to them when these dear beings are no more." Having decided that one must make a choice between finding God's or humanity's agency in child mortality, Chambers chose human agency. Only then, he argued, could action be directed toward making things better in future. Critics argued that the loss of God's presence in human lives would destroy all sense of moral responsibility and result in social chaos. But Chambers argued that moral responsibility would be *heightened* once individuals understood that their welfare depended upon learning and obeying natural laws in the here and now, rather than focusing on heaven in the hereafter.[40]

When *Vestiges* appeared in 1844 Darwin had been working away at his own theory of evolution for almost a decade. Like Chambers, he had included human beings within that system from the beginning.

Throughout 1845, he paid close attention to professional naturalists' efforts to ridicule Chambers's evolutionary "Law of Development" into oblivion. He read Sedgwick's eighty-eight-page review, including the lines in which his mentor confessed he first thought the author must be a woman because no *man* could have written so unscientific a book! The shocked, angry responses from Sedgwick and others helped convince Darwin to put his own two-hundred-page manuscript safely away on a shelf. Meanwhile, his good friend Hooker was complaining about French naturalists for speculating about species boundaries without having done any real taxonomic work. Darwin took the complaint personally: "How painfully (to me) true is your remark that no one has hardly a right to examine the question of species who has not minutely examined many." So between 1846 and 1854 he completed a meticulous study of barnacles, earning his spurs as an expert taxonomist (his barnacle monograph is still in use today). No one, by the time he was done, could accuse him of not having "minutely examined many" species, should he decide to publish on their origins.[41]

This Very Old Argument

As Darwin began his work on barnacles, mothers and fathers across the Irish Sea were experiencing child loss at a famine-level scale. A blight caused by the water mold *Phytophthora infestans* had destroyed the potato crop upon which much of the Irish population depended for sustenance. The Great Famine began in 1845 and lasted until 1849. By the time it was over, a million Irish had perished from starvation, typhus, and cholera, and a million more had emigrated to the United States and the farthest reaches of the British Empire.

As rumors of famine spread, Darwin took time away from his barnacles to send some notes on the effect of the blight on his own plants to Professor Henslow. He also corresponded with Miles J. Berkeley, who was arguing that a pathogenic organism caused the blight, rather than bad weather, the soil, or Irish immorality. In view of the poor's dependence upon the potato crop, Darwin agreed with Henslow's suggestion that "gentlefolk" not buy potatoes. "The poor people," Darwin wrote, "wherever I have been, seem to be in great alarm: my labourer here has not above a few weeks consumption & those not sound." He pondered whether new potato seed from South America would halt the disease (his grandfather had proposed that crossing plants yielded healthier progeny).[42]

As people tried to make sense of the famine and figure out what to do, Henslow thought "it seems to be a providential arrangement" that parts of the potato seemed uninjured, and might be ground to a wholesome flour (he was wrong that the flour was edible; it, too, rotted almost immediately). Others cited "Malthusian laws" and located the origin of the famine within Providential arrangements of the world. (Clearly, despite Chambers's arguments, better knowledge of natural laws did not automatically lead to action to ameliorate suffering; indeed, in this case the powers-that-be called upon nature's laws to do just the opposite.) Meanwhile, the famine flamed the fires of discontent and helped to inspire the failed European Revolutions of 1848, revolutions amid which Marx and Engels would take on Malthus's vision of nature, and the industrialist-capitalist system it justified, at full tilt.[43]

With the exception of Hooker, Darwin kept his theorizing that the same laws that determined the vitality of potatoes and the mortality of animals also governed human populations to himself. There was so much work to do and so many questions to be answered. Meanwhile, the Darwin family grew. By 1850 three more children had arrived in the Darwin home. Eleven-year-old William was the eldest, then Anne (nine), Henrietta (seven), George (five), Elizabeth (three), and Francis (two), and the newest baby was Leonard. The butler, Parslow, grumbled that there were sometimes "twenty pair of little shoes to clean."[44]

Although the Darwins had no chance of their children starving to death, wealth and privilege could not protect families from the infectious diseases that plagued childhood. As Darwin tried to develop a consilient theory that would explain the world around him, his beloved ten-year-old Annie Darwin died on April 23, 1851. Annie's sister Henrietta later recounted how, when Annie was nine, her health had begun to break down. Darwin eventually sent her to Malvern with her sister, nurse, and governess so she could try the water cure under the supervision of Darwin's own doctor, Dr. Gully. When Annie grew worse, Darwin went to Malvern too. "I well remember his arrival," Henrietta later wrote, "and how he flung himself on the sofa in an agony of grief." Expecting another baby, Emma had stayed at Down House. Letters went back and forth between the day of Darwin's arrival and Annie's death six days later. "You would not in the least recognize her with her poor hard, sharp pinched features," Darwin wrote. He knew complete honesty was the only option in writing to Emma, and he confessed: "I could only bear to look at her by forgetting our former dear Annie."[45]

He wrote several times a day. "Poor Annie has just said 'Papa' quite distinctly," he wrote on the Saturday: "Our poor child has been fearfully ill, as ill as a human being could be, it was dreadful that night the Dr told me it would probably be all over before morning." Soon, he had to write that the end had indeed come. "We must be more and more to each other, my dear wife" he wrote, now that Annie was gone. He felt desolate thinking of her frank, cordial manners. "Our poor dear dear child has had a very short life," he added, "but I trust happy, and God only knows what miseries might have been in store for her." This was the only solace he had offered to William Owen a few years earlier: Given his deep awareness of the potential for great suffering in human life, he took comfort in the knowledge that the chance of Annie experiencing additional pain was now gone.[46]

Darwin wrote of Annie's "poor dear little soul" and her perfect "gentleness, patience and gratitude," even as she lay dying. But he did not write of heaven. He did not find solace, as William Owen did, in the belief that "all is for the best & that the blow has been struck in mercy by that Almighty Spirit who we are bound to believe cannot err, & without whose knowledge & will not a Sparrow falls." In return, Emma did not write of such things to him. "My only hope of consolation," she wrote instead, "is to have you safe home and weep together." She added: "You do give me the only comfort I can take, in thinking of her happy, innocent life."[47]

Henrietta Darwin later wrote of the influence Annie's death had on their mother as follows: "It may almost be said that my mother never really recovered from this grief. She very rarely spoke of Annie, but when she did the sense of loss was always there unhealed. My father could not bear to reopen his sorrow, and he never, to my knowledge, spoke of her." Charles Darwin did pour his grief onto paper in private, in a loving account of Annie's final illness: "Her conduct in simple truth was angelic. She never once complained; never became fretful; was ever considerate of others, and was thankful in the most gentle pathetic manner for everything done for her. . . . When I gave her some water, she said, 'I quite thank you'; and these, I believe, were the last precious words ever addressed by her dear lips to me." They had lost, he concluded, "the joy of the household, and the solace of our old age. . . . Oh, that she could now know how deeply, how tenderly, we do still and shall ever love her dear joyous face! Blessings on her!"[48]

Darwin was forty-two years old when Annie died. Thirty years later, while debating Edward Aveling (the atheist son-in-law of Karl Marx) over whether unbelievers should aggressively try to convert others, Darwin mentioned that he "never gave up Christianity until I was forty years of age." Taken literally, that would place his conversion to unbelief in 1849, two years prior to Annie's death. Taken more generally (perhaps Aveling forgot an "around forty"), one might forward the date two years. The timing, in any case, was close.[49]

Scholars disagree regarding the role of Annie's death in moving Darwin toward unbelief. But it is difficult to read Darwin's explanations of his changing religious views (composed in the 1870s) without Annie's death appearing between the lines. Take the following passage, included in a long list of factors that inspired his gentle descent toward unbelief:

That there is much suffering in the world no one disputes. Some have attempted to explain this with reference to man by imagining that it serves for his moral improvement. But the number of men in the world is as nothing compared with that of all other sentient beings, and they often suffer greatly without any moral improvement. A being so powerful and so full of knowledge as a God who could create the universe, is to our finite minds omnipotent and omniscient, and it revolts our understanding to suppose that his benevolence is not unbounded, for what advantage can there be in the suffering of millions of the lower animals throughout the almost endless time? This very old argument from the existence of suffering against the existence of an intelligent First Cause seems to me a strong one.[50]

Darwin seems to be talking mainly of nonhuman animals (i.e., other sentient beings), but, of course, by the time he wrote these lines he had completely collapsed the boundary between humans and animals. That said, Darwin did not need Annie's death to repudiate William Owen's belief that God attended to the fall of every sparrow. He had been asking questions about suffering, including why differential mortality existed between the children of Indigenous and European populations, for nearly two decades by the time Annie died. Furthermore, a righteous drift away from belief in the all-good, all-wise, all-knowing God of Christianity was all around him: in his brother Erasmus's salon; in the writings of his grandfather Erasmus; and in the radical, "freethinking" books of some of his brother's friends. Indeed, earlier that year Harriet Martineau had published (with Henry George Atkinson) a book entitled *Letters on the Laws of Man's Nature and Development* that contained Martineau's public confession of unbelief (or, as she called it, necessarianism). Like the author of *Vestiges*, Martineau insisted that true solace arose from focusing on the study of "necessary cause and effect" (natural laws) as the means of both understanding suffering and ameliorating it in future. Unlike *Vestiges* (which offered its readers at least a vague hope that heaven exists), she called faith in heaven an irrational, misguided, and selfish desire for eternal life that caused more misery (especially when accompanied by belief in hell) than it was worth.[51]

Darwin may have had a copy of Martineau's book with him in Malvern when Annie died. The next day, he asked Fanny Mackintosh to "take Miss Martineau to Erasmus." (He did not record completing the book until August 4, 1852.) We know, however, that he was sympathetic with at least some components of Martineau's "necessarian" vision of the world. Martineau had fully embraced, and then translated, the work of a French philosopher named August Comte, whose work also captivated Darwin. Comte famously argued that history proceeded through a "Law of Three Stages," each determined by a particular method of obtaining knowledge. In the theological stage, humanity obtained knowledge via

Scripture and authority, and relied on supernatural explanations. In the metaphysical stage, explanations were rooted in abstractions disconnected from human experience (for example, ideas like liberty, equality, and fraternity). In the final, positive stage (positive as in true, rather than false), humans relied solely on naturalistic, scientific explanations for knowledge of nature, themselves, and society. (Upon first reading of Comte's work, Darwin jotted "Zoology itself is now purely theological" in his notebook. This was not a compliment to zoologists.) Meanwhile, Comte coined words to capture what the "positive" stage of society must be based upon: words like *altruism* (the belief that men must serve each other, rather than focus on God or heaven), and *sociology* (the scientific study of society in the name of rational reform).[52]

Darwin was drawn to explanations via natural laws for several reasons. The search for secondary causes certainly gave ambitious naturalists like himself something to do, beyond just documenting new species and purposeful parts. But naturalistic explanations also provided an alternative means of understanding human experience. "I am inclined," he wrote, almost a decade after Annie's death, "to look at everything as resulting from designed laws, with the details, whether good or bad, left to the working out of what we may call chance." Death and the various afflictions to which individuals were subject came to mind when he grasped for examples. "The lightning kills a man," he imagined, "whether a good one or bad one, owing to the excessively complex action of natural laws,—a child (who may turn out an idiot) is born by action of even more complex laws,—and I can see no reason, why a man, or other animal, may not have been aboriginally produced by other laws." He briefly entertained the idea that "all these laws may have been expressly designed by an omniscient Creator, who foresaw every future event & consequence," but quickly added that "the more I think the more bewildered I become." He tended to end such thoughts abruptly: "I feel most deeply that the whole subject is too profound for the human intellect," he added. "A dog might as well speculate on the mind of Newton." Six years later, when a correspondent asked him whether he thought natural selection consistent with belief in a personal, infinitely good God, he replied: "It has always appeared to me more satisfactory to look at the immense amount of pain

& suffering in this world, as the inevitable result of the natural sequence of events, i.e., general laws, rather than from the direct intervention of God."[53]

As Charles drifted away from belief in the God who attended to the fall of every sparrow, Emma held onto Unitarianism. Ultimately, by the time Charles died they had developed very different, indeed irreconcilable, conceptions of what Christianity even is. We know this because of an exchange between Emma and her son, Francis, as they prepared Darwin's account of his "Religious Beliefs" for publication after his death. Darwin had clearly written this informal "autobiography" solely for his own family, but in the wake of stories circulating among evangelicals that he had converted to Christianity on his deathbed, and among atheists that he died an atheist, the family had decided to publish Darwin's own words on the matter. Emma and Francis agreed to publish most of Darwin's meandering account of his changing religious beliefs. Emma approved the inclusion of Darwin's heartfelt paragraph about the problem of suffering and the lines insisting that the discovery of natural laws undermined belief in miracles. But she firmly opposed printing her husband's following words:

I can indeed hardly see how anyone ought to wish Christianity to be true, for if so the plain language of the text seems to show that the men who do not believe, and this would include my Father, Brother, and almost all of my friends, will be everlastingly punished.

And this is a damnable doctrine.[54]

Emma had the following to say about Charles's interpretation of the "plain language of the text": "I should dislike the passage in brackets to be published. It seems to me raw. Nothing can be said too severe upon the doctrine of everlasting punishment for disbelief—but very few now would call that 'Christianity.'"[55]

Of course, Victorians had been debating what precisely constituted Christianity for some time. Anglican orthodoxy was quite clear on the fate of the nonbeliever. In the words of Pearson's *An Exposition of the*

Creed, no one "shall ever escape eternal flames, except we obtain the favour of God before we be swallowed by the jaws of Death."[56] Darwin had clearly rebelled from the idea that men like his father, brother, and best friends would be damned for not believing in Christianity. He had plenty of company in his rebellion. The philosopher Francis Newman, whose books Darwin read, insisted that surely truly progressive and ethical men dispensed with such barbaric doctrines. For many dissenters, the "otherworldly salvation motif" of orthodox Anglicanism stood in stark contrast with what historian Howard Murphy has called the age's meliorist belief that "the life of man on this earth both can and should be progressively improved through sustained application of human effort and intelligence." Indeed, although the Victorian "crisis of faith" appears in retrospect as inspired by geological discoveries and evolution, upon closer examination, new visions of the past filled gaps first created by ethical rebellions against orthodox doctrines.[57]

Emma agreed with this rebellion against belief in eternal damnation. Unitarians had worked hard to adjust to the times, in order to ensure that individuals could dispense with certain doctrines yet not abandon Christ's message of salvation altogether. By the 1850s even some Anglicans were arguing that belief in eternal punishment must be given up. But for some, including Darwin, it was all too little too late. Besides, he had found a different set of explanations for the world. When, in writing his autobiography, he cited the problem of suffering as influencing his views, he added that the existence of much suffering agreed well with the view that all organic beings had been developed through natural selection.[58]

By the spring of 1856 some of Darwin's closest friends were learning how much of the old links between Christianity and nature he had given up. As a Unitarian, Charles Lyell was quite comfortable with driving natural law into the deep past. He had helped expand the age of the earth beyond what even Sedgwick (who was no biblical literalist) was willing to countenance. Lyell knew Darwin was working on the "species problem," but he did not know about his "theory by which to work" until Darwin showed him the pigeon lofts. Darwin walked through his argument as pigeons cooed and flew about: First, despite profound differences between pigeon breeds, ornithologists knew they all descended from the common rock pigeon. Yet any naturalist, taking them as wild birds, would surely classify pigeon breeds (fantails, pouters, tumblers, Jacobins, barbs, carriers, and more) as at least three different genera and about fifteen *species*. Surely one might extend this lesson, Darwin insisted, to assessing the possibility that different species of finches on the Galapagos Islands might in fact have arisen from one common ancestor, long ago. Second, most of the time pigeon breeders did not have a particular purpose or goal in mind for a trait: they simply bred from their "best birds" and thus, over time, slowly altered the breeds. That, Darwin argued, was how natural selection worked. "Adaptations" were the result of a purely mechanical process of variation, differential survival rates, and the higher reproductive success of certain populations within an incessant struggle for existence. The Divine Watchmaker, who paid such close attention to the design of the swallow's wing that one could trust He also attended to the joys and afflictions of one's own life, was gone.

After Lyell's visit, Darwin reported with triumph to an American geologist who knew he was working on "the species problem" that "It may sound presumptuous, but I think I have to a certain extent staggered even Lyell." He had indeed. Having ridiculed Lamarck twenty years earlier as unscientific and dangerous, Lyell went home and opened a set of notebooks in which he wrestled with the prospect that Darwin was right. He saw the weight of Darwin's arguments. Darwin had a purely uniformitarian explanation, rooted in the struggle for existence, a phenomenon that Lyell himself had given as an explanation for extinction. Furthermore, Lyell had provided the massive amount of time required for natural selection to work. He had also (inadvertently) provided an explanation for why, if Darwin was right, no transitional fossils could be found: in arguing against Sedgwick's "theory of successive, progressive creations" Lyell had insisted that the lack of, say, mammals, in the lowest levels of the fossil record could not be taken as proof that such creatures had not roamed the earth long ago.[59]

For the next few years, even as he urged Darwin to publish, Lyell struggled with whether, as he put it, to "go the whole Orang" with Darwin. Accept transmutation, he wrote in his notebook in June 1859, and "the Garden of Eden, Milton's Paradise, the Golden Age, all vanish. In its place we have an ancestry of tens of thousands of years as unprogressive as the Elephants or Orang tribes." Then he wrote down something the great physicist Michael Faraday had said in his lecture on "The Education of Judgment": "The point of self-education" was, Faraday said, to "teach the mind to resist its desires & inclinations, until they are proved to be right." A few months later Lyell jotted down a statement by John Stuart Mill: "The person who has to think more of what an opinion leads to than of what is the evidence of it, cannot be a philosopher or a teacher of philosophers." He was trying his hardest to give Darwin's ideas a fair hearing. But he was, indeed, staggering beneath the potential costs of giving up his firm belief in the dignity of man.[60]

Alone at his desk, accompanied solely by his own hopes and anxieties, Lyell confessed that there was certainly one point on which the evolutionary creation of species appealed: if Darwin was correct, then one must no longer labor under a doctrine of the direct agency of the Creator "as the immediate source of individuality," which "labours under the difficulty that abortion, inferior, diseased, immoral, stupid, insane creatures are not the result of laws but of special intervention." Lyell saw that if Darwin's theory applied to the origin of man, then ideas about God's governance must be altered. He saw the benefits. But he was not sure they were worth the costs.[61]

No matter where the idea led, however, Lyell knew his friend had a strong case. Despite his profound misgivings, he urged Darwin to publish or risk being scooped. In response, Darwin groaned to Hooker that he hated the idea of publishing "a mere sketch or outline of future work in which full references &c sh^d. be given.—Eheu, eheu, [alas, alas] I truly believe I sh^d. sneer at anyone else doing this, & my only comfort is, that I *truly* never dreamed of it, till Lyell suggested it, & seems deliberately to think it adviseable." He feared it was unscientific to publish results without the full details that had led to those results, and delayed. "I begin *most* heartily to wish," he lamented, "Lyell had never put this idea of an Essay into my head."[62]

Why Do Some Die and Some Live?

Over a decade earlier, as Darwin had worked away at his theory in the late 1840s, a young man named Alfred Russel Wallace was also reading *Vestiges of the Natural History of Creation*, Thomas Malthus's *Essay on Population*, and Charles Lyell's *Principles of Geology*. In the 1850s, Wallace was traveling as a naturalist in the tropics studying the variation and geographical distribution of animals, plants, and peoples. In 1858, sick with malaria in the Malay Archipelago and reflecting on the enormous destruction of animal life that arose from the limits of subsistence, Wallace asked himself: "Why do some die and some live?" He, too, found the answer in the struggle for existence and a higher rate of survival among varieties that had some advantage in that struggle. He, too, thought these pressures might drive forms across the species boundary, and perhaps even create new purposeful parts and organs.

Wallace wrote up his idea in an essay and sent it to Darwin, completely unaware that the famous naturalist of the voyage of the *Beagle* had been thinking along similar lines for two decades. As a man with few connections, Wallace was asking Darwin, with whom he had on occasion corresponded, to forward his essay to Charles Lyell. Four months after they were posted, Wallace's letter and manuscript arrived at Down House. Darwin was stunned. He wrote immediately, as requested, to Lyell: "I never saw a more striking coincidence. If Wallace had my MS sketch written out in 1842 he could not have made a better short abstract!" He included Wallace's essay, adding with profound understatement: "It seems to me well worth reading." He then preempted Lyell's "I told you so" with the words: "Your words have come true with a vengeance that I should be forestalled." He added that he would of course write Wallace and offer to send the essay to any journal. With those words Darwin gave up all claim of priority to the idea that had governed his days, thoughts, and ambitions for twenty years.[63]

Who was this naturalist who had hit upon the same idea as Darwin? In many ways, Alfred Russel Wallace was from a different world. Born into a middle-class family that had fallen on tough times, he had to work for a living. He traveled the world and collected specimens to earn money, rather than for amusement. When he traveled to South America (two decades after Darwin) he had to keep track of every pence and calculate the worth of every bird or mammal he shot and every insect he pinned.

But Wallace and Darwin also had much in common. As a boy Wallace had attended the Anglican Church of his parents. On Sundays only the Bible, *Pilgrim's Progress*, or *Paradise Lost* were allowed. Then, as a teenager, he began questioning orthodox doctrine amid both observing the injustices in the society around him and listening to lectures at the "Hall of Science," a club for workers founded by socialist reformers like Robert Owen. There, "a pervading spirit of skepticism, or freethought as it was then called . . . strengthened and confirmed my doubts as to the truth or value of all religious teaching." The radical German theologian David Strauss's *Life of Jesus* made a strong impression. Strauss's book convinced Wallace that the miracles in the Gospels were "mere myths, which in periods of ignorance and credulity always grow up around great men," while arguments against the doctrine of eternal punishment convinced him that orthodox Christianity was "degrading and hideous." He knew a lady, he recalled years later, who tried to commit suicide after listening to a sermon on this doctrine.[64]

Years later, he reflected on the alternative explanations of suffering he learned as a young man:

> It must have been in one of the books or papers I read here [at Owen's "Hall of Science"] that I met with what I dare say is a very old dilemma as to the origin of evil. It runs thus: "Is God able to prevent evil but not willing? Then he is not benevolent. Is he willing but not able? Then he is not omnipotent. Is he both able and willing? Whence then is evil?"[65] This struck me very much, and it seemed quite unanswerable. . . . If the argument did not really touch the question of the existence of God, it did seem to prove that the orthodox ideas as to His nature and powers cannot be accepted.[66]

From an early age, influenced by the grand campaigns for reform by Robert Owen and others, Wallace's criteria for the right kind of religion was set: the "true and wholly beneficial religion was that which inculcated the service of humanity and whose only dogma was the brotherhood of man." He later wrote that, not having found such a religion, the result of all his reading, listening, and thinking was that as a young man he could best be described "by the modern term 'agnostic.'"[67]

Though trained as a land surveyor, Wallace was also an ardent naturalist. The radical thinking described above meant that he did not agree with some of the fundamental theological assumptions to which the dominant explanation of the origin of species had been tied—namely, that a close, personal God had independently created all animal and plant forms. He clearly had little patience with the theory of special creation (or, as he called it, the "continual interference hypothesis"). Indeed, he ridiculed those who found evidence of God's benevolent design everywhere. Naturalists described, for example, the soft scar of the coconut as a wise "contrivance" that allowed the embryo to emerge. "What should we think,' Wallace demanded with disdain, "if as proof of the superior wisdom of some philosopher, it was pointed out that in building a house he had made a door to it. . . . But this is the kind & degree of design imputed to the Deity as a proof of his infinite wisdom." Wallace despised the arrogance of those who assumed that by producing such accounts of creation, they became "well acquainted with the motives of the Creator." But it was the use of the theory of special creation to justify the status quo that was, to Wallace, the most damning mark against it. (As we will see, he eventually became a socialist and campaigned for radical land reform, while writing passionately of injustices such as the "sufferings of so many infants needlessly massacred through the terrible defects of our vicious social system.")[68]

Meanwhile, inspired by his reading of Darwin's *Journal of Researches* and Alexander von Humboldt's *Personal Narrative of Travels to the Equinoctial Regions of America*, in 1847 Wallace cooked up a scheme with fellow naturalist Henry Walter Bates to go to the Amazon. Wallace hoped to gather facts "towards solving the problem of the origin of species," a subject he and Bates had often discussed. The two working-class naturalists would pay their way by collecting specimens to sell to wealthy collectors and museums on their return home. In correspondence just prior to their adventure, Wallace confessed that, unlike Bates, he did not consider *Vestiges* to be a hasty generalization "but rather as an ingenious hypothesis strongly supported by some striking facts and analogies but which remains to be proved by more facts & the additional light which future researches may throw upon the subject." He thought the book furnished a wonderful "incitement to the collection of facts & an object to which to apply them when collected," a view that would influence his time in the field, with profound results.[69]

Wallace's life in the field was marred by both tragedy and triumph. He was plagued by dysentery (severe diarrhea) and frequent bouts of malaria, which he dealt with by the usual methods of purges and cathartics. Mosquitoes were a constant nuisance. At some sites his collecting was marred by biting flies, where "the torments I suffered when skinning a bird or drawing a fish, can scarcely be imagined by the unexperienced." His much-loved younger brother, Herbert, who came out to try his hand at collecting, died of yellow fever. Then Wallace, returning home with a collection that would have vied with the best in Europe, lost two years' hard work and almost his life when the ship caught fire and sank. "How many weary days and weeks had I passed," he lamented, "upheld only by the fond hope of bringing home many new and beautiful forms from those wild regions . . . (all would) prove that I had not wasted the advantages I had enjoyed, and would give me occupation and amusement for many years to come! And now everything was gone."[70]

For a naturalist, the loss of his collection was an affliction indeed. By this time in his life, Wallace had long since abandoned orthodox means of explaining such loss. He saw it as neither a test of faith nor punishment for pride and worldliness. Forcing himself to think on the state of things "which actually existed" rather than what might have been, and unable to call on either family fortune or a rich wife's dowry, he decided to take his chances on another voyage. In 1854, after a brief break in England, he left for the Malay Archipelago.

Meanwhile, Wallace pondered the question of the origin of species. He clearly found *Vestiges*'s naturalistic approach to the origin of species appealing, but he also knew it was extremely speculative. As a good field naturalist, he paid close attention to the questions a new theory of the origin of species would have to answer, especially ones that concerned taxonomy and geographical distribution. "How and why," he asked, "do they change into new and well-defined species, distinguished from each other in so many ways; why and how do they become so exactly adapted to distinct modes of life; and why do all the intermediate grades die out (as geology shows they have died out) and leave only clearly defined and well-marked species, genera, and higher groups of animals?"[71]

He found his answer while suffering from a bout of malaria. For, having nothing to do "but to think over any subjects then particularly interesting me," he recalled Malthus's "Principles of Population," which he had read more than a decade earlier. Convinced by Malthus that inevitable checks on population increase (disease, accidents, war, and famine) operated in both man and animals, Wallace realized that "the destruction every year from these causes must be enormous in order to keep down the numbers of each species, since they evidently do not increase regularly from year to year, as otherwise the world would long ago have been densely crowded with those that breed most quickly." He continued:

> *Vaguely thinking over the enormous and constant destruction which this implied, it occurred to me to ask the question, Why do some die and some live? And the answer was clearly, that on the whole the best fitted live. From the effects of disease the most healthy escaped; from enemies, the strongest, the swiftest, or the most cunning; from famine, the best hunters or those with the best digestion; and so on. Then it suddenly flashed upon me that this self-acting process would necessarily improve the race, because in every generation the inferior would inevitably be killed off and the superior would remain—that is, the fittest would survive.[72]*

When Wallace asked the question, "Why do some die and some live?" the problem was not abstract: his brother had died; he had lived. Now, suffering from malaria, he might die any day. The same question had plagued countless others, including Darwin. Ultimately, both Wallace and Darwin found their answer to the question by combining their curiosity as naturalists about the origin of species with the claims of a parson who had tried to explain why a benevolent God allowed so much suffering in the world.

After Wallace's letter and essay arrived, Charles Lyell and Joseph Dalton Hooker convinced Darwin to allow them to send the essay he had written in 1844 and a letter to American botanist Asa Gray from 1857 in which he had outlined his theory, along with Wallace's essay, to the Linnean Society of London. In their letter accompanying these writings, Lyell and Hooker informed the society that the papers reflected work pursued "independently and unknown to one another." They also noted Darwin's reluctance to forward anything but Wallace's paper, and how they had explained to him that it would be in "the interests of science generally" for his views, "founded on a wide deduction from facts, and matured by years of reflection," to be placed before "the scientific world."[73]

Debates regarding who deserves credit for natural selection and whether Darwin dealt with Wallace fairly have raged for generations. Few mention the fact that, as historian Janet Browne notes, "Darwin was in no state of mind to make any kind of balanced assessment about competing claims over his or Wallace's priority." A day after Wallace's letter arrived, fourteen-year-old Henrietta Darwin came down with diphtheria, one of the leading causes of death for both children and adults. Known as the "strangling angel of children," this painful illness produced a wing-shaped, white membrane that grew into the nose, esophagus, or larynx and sometimes closed the airway entirely. That same day, news arrived from twelve-year-old George's boarding school that he had the measles. Then, two days after Darwin first read Wallace's letter, his eighteen-month-old baby Charles fell dangerously ill.[74]

As letters from Lyell and Hooker arrived at Down House trying to convince Darwin that it would not be dishonorable to forward his own essay with Wallace's manuscript, baby Charles grew worse. It was scarlet fever, the same illness that had killed Robert Chambers's little girl, Margaret, and one of the most feared childhood killers of the nineteenth century. Parents watched helplessly as the sudden onset of a painful sore throat was followed by vomiting, chills, and high fever, and then the bright red rash that gave the illness its name. "Scarlatina" was acute, contagious, and thought to sometimes cause permanent hearing loss. Physicians described the worst cases descending into convulsions, delirium, and death. And it all happened with such cruel swiftness. The American physician Caspar Morris, reflecting on twenty-five years' observation of the disease, wrote: "No disease had so unexpected or violent an onset. The parent sits down in the evening in the happy centre of a group of smiling objects of affection, his heart swelling with delightful anticipation as his eye glances around the circle; and ere the next return of the same weekly period, half of them slumber in the embrace of death."[75]

Attempts to apply lessons learned from smallpox and prevent the worst ravages of the disease through vaccination failed. Although many physicians believed that the disease spread by contagion (as opposed to atmospheric changes), no one knew what caused it. Germ theory, which would provide the framework for identifying the real culprit (a bacteria called *Streptococcus pyogenes*), was more than a generation away.

When little Charles fell ill, three babies in the village of Downe had already died from the disease. A week later Emma and Charles lost a child for the third and final time. The baby's terrible final days are hinted at in the brief letter Darwin wrote to Hooker the day after it was all over: "I hope to God he did not suffer so much as he appeared. . . . It was the most blessed relief to see his poor little innocent face resume its sweet expression in the sleep of death." Later that day he wrote again after Hooker requested he send copies of his old essays so they could be forwarded to the Linnean Society with Wallace's manuscript: "I am quite prostrated & can do nothing but I send Wallace & my abstract of letter to Asa Gray, which gives most imperfectly only the means of change & does not touch on reasons for believing species do change. I daresay all is too late. I hardly care about it."[76]

The set of documents gathered by Hooker and Lyell, including Wallace's essay, were read at the Linnean Society two days later, on July 1. In introducing the papers, Hooker and Lyell explained that they all related to "the Laws which affect the Production of Varieties, Races, and Species."[77] In other words, the documents announced the Darwin-Wallace theory of evolution via natural selection to the scientific world.

In contrast to *On the Origin of Species*, the 1844 essay included in the Darwin-Wallace papers had not been meant for publication. Its description of nature is bleaker and harsher than *Origin*, and arguably closer to Darwin's own view, first forged in the suffering witnessed on the voyage, and then at the bedsides of his own babies: "Nature may be compared to a surface on which rest ten thousand sharp wedges touching each other and driven inwards by incessant blows." (In *Origin*, Darwin tempered these words with the reflection that, "when we reflect on this struggle, we may console ourselves with the full belief, that the war of nature is not incessant, that no fear is felt, that death is generally prompt, and that the vigorous, the healthy, and the happy survive and multiply.")[78]

Darwin conceded that to fully realize the extent of the struggle for existence "much reflection is requisite." Malthus on man should be studied, and non-native cattle and horses when first introduced to South America. "Reflect on the enormous multiplying power inherent and annually in action in all animals," he advised. "Reflect on the countless seeds scattered by a hundred ingenious contrivances, year after year, over the whole face of the land; and yet we have every reason to suppose that the average percentage of each of the inhabitants of a country usually remains constant." Could it be doubted, Darwin urged, that

in the consequent struggle of each individual to obtain subsistence, that any minute variation in structure, habits, or instincts, adapting that individual better to the new conditions, would tell upon its vigour and health? In the struggle it would have a better chance of surviving; and those of its offspring which inherited the variation, be it ever so slight, would also have a better chance. Yearly more are bred than can survive; the smallest grain in the balance, in the long run, must tell on which death shall fall, and which shall survive.[79]

Read before a specialized scientific society, the Darwin-Wallace papers made little impact. So, over the next year, Darwin composed an "Abstract" of his long-envisioned book on species. He entitled it *On the Origin of Species by Natural Selection, or the Preservation of Favoured Races in the Struggle for Life*. The book appeared in London bookstalls on November 24, 1859. As historian Janet Browne notes, Darwin "faced the arduous task of reorienting the way Victorians looked at nature" away from Paley's "It is a Happy world after all." But of course, some did not need to reorient anything: for some, Darwin's nature mapped onto their experience of the world, and British society, quite well.[80]

CHAPTER SIX

Grandeur in This View of Life

UPON OPENING *ON THE ORIGIN OF SPECIES*, READERS FOUND TWO QUOTA-tions Darwin selected to appear opposite the title page. First, the words of the philosopher of science Reverend William Whewell: "To conclude, therefore, let no man out of a weak conceit of sobriety, or an ill-applied moderation, think or maintain, that a man can search too far or be too well studied in the book of God's word, or in the book of God's works; divinity or philosophy; but rather let men endeavour an endless progress or proficience in both." Second, those of the seventeenth-century philos-opher Francis Bacon: "But with regard to the material world, we can at least go so far as this—we can perceive that events are brought about not by insulated interpositions of Divine power, exerted in each particular case, but by the establishment of general laws."

Both quotations appealed to British science's strong tradition of studying secondary causes while acknowledging the first cause of all phenomena, God. The choices were highly strategic. After all, Darwin was proposing an alternative, mechanistic explanation for phenomena that, for generations, the theory of special creation concluded could only be explained via the direct intervention of the first cause. So he began by reminding his readers that to explain things via secondary causes did not necessarily call God's governance into question.

There is no evidence that his inclusion of either quote, at least in 1859, was somehow insincere. Although by this point in his life Darwin had clearly abandoned Christianity, he still found it extremely difficult to imagine "this immense and wonderful universe, including man with his capacity of looking far backwards and far into futurity, as the result of blind chance or necessity." He thus felt "compelled to look to a First Cause having an intelligent mind in some degrees analogous to that of man. . . . This conclusion was strong in my mind about the time, as far as I can remember, when I wrote the *Origin of Species*."[1]

Victorian readers could still find William Paley and Thomas Malthus's law-governing God in Darwin's vision of nature. Finding Christianity's close personal God "without whose knowledge & will not a Sparrow falls" could prove more difficult, though not, as we will see, impossible. Whether that difficulty reflected a virtue or vice of *On the Origin of Species* was soon up for debate. Taking three "Darwinians" in turn (Darwin himself, his American supporter Asa Gray, and his British defender Thomas Henry Huxley), we can see that the problem of suffering, including child loss, lay between the lines of these decisions. Sometimes it rose quite close to the surface.

WE NEED NOT MARVEL

Darwin knew that the theory he was proposing could not "be directly proved." Rather, as he wrote to Hooker, "the doctrine must sink or swim according as it groups and explains phenomena." The view that evolution must be judged by whether it explained the most facts reflected William Whewell's doctrine of consilience (i.e., the strongest theory is the one that explains the greatest amount of seemingly disparate phenomena). In a brief introduction, Darwin emphasized the wide range of puzzles that faced working taxonomists, puzzles he would, by the end of the book, argue that evolution explained:

> In considering the Origin of Species, it is quite conceivable that a naturalist, reflecting on the mutual affinities of organic beings, on their embryological relations, their geographical distribution, geological succession, and other such facts, might come to the conclusion that each species had not been independently created, but had descended, like varieties, from other species.[2]

Darwin then briefly summarized his explanation of how, if he was right, purposeful parts, new organs, and remarkable co-adaptations had arisen. Thanks to Malthus, his readers already knew about the "struggle for existence." Darwin had to convince them that, given the tremendous intensity of that struggle, "any being, if it vary however slightly in any manner profitable to itself, under the complex and sometimes varying conditions of life, will have a better chance of surviving, and thus be *naturally selected*." Most importantly, he had to convince his readers that this secondary cause could push varieties across the species boundary, and that what he called "descent with modification via natural selection" explained the origin of both species and purposeful parts better than the theory of special creation.[3]

Throughout the book, Darwin countered William Paley's vision of joyous insects flitting about a "happy world after all" with a stern demand that his readers see nature *as it is*: "We behold the face of nature bright with gladness," he wrote, "we do not see, or we forget, that the birds which are idly singing round us mostly live on insects or seeds, and are thus constantly destroying life; or we forget how largely these songsters, or their eggs, or their nestlings, are destroyed by birds and beasts of prey." In fact, he wrote, every single organic being "lives by a struggle at some period of its life." Heavy destruction "inevitably falls either on the young or old, during each generation or at recurrent intervals." Passages like this portrayed nature as a violent, competitive place in which each organic being "at some period of its life, during some season of the year, during each generation or at intervals, has to struggle for life, and to suffer great destruction." The passage from the 1844 essay comparing nature to "ten thousand sharp wedges" remained with just slight revision: "The face of Nature may be compared to a yielding surface, with ten thousand sharp wedges packed close together and driven inwards by incessant blows, sometimes one wedge being struck, and then another with greater force."[4]

Certain that all of Paley's most famous examples of benevolently designed, purposeful parts would be in the minds of his readers, Darwin then argued that "descent with modification" explained things that Paley's explanation of species could not:

How strange it is that a bird, under the form of woodpecker, should have been created to prey on insects on the ground; that upland geese, which never or rarely swim, should have been created with webbed feet; that a thrush should have been created to dive and feed on sub-aquatic insects. But on the view of each species constantly trying to increase in number, with natural selection always ready to adapt the slowly varying descendants of each to any unoccupied or ill-occupied place in nature, these facts cease to be strange, or perhaps might even have been anticipated.[5]

Other facts—facts that had troubled more than just naturalists—ceased "to be strange" as well. In a chapter on instincts, Darwin wrote:

It may not be a logical deduction, but to my imagination it is far more satisfactory to look at such instincts as the young cuckoo ejecting its foster-brothers,—ants making slaves,—the larvæ of ichneumonidæ feeding within the live bodies of caterpillars,—not as specially endowed or created instincts, but as small consequences of one general law, leading to the advancement of all organic beings, namely, multiply, vary, let the strongest live and the weakest die.[6]

Darwin was not trying to make cuckoos, slavery, or the Ichneu-monidae good. He assumed his readers would agree each caused great suffering. Based on that assumption, he offered an alternative explanation of why such things existed. Accept descent with modification via natural selection, Darwin argued, and we need not marvel "if all the contrivances in nature be not, as far as we can judge, absolutely perfect; and if some of them be abhorrent to our ideas of fitness." In doing so he dispensed with the need to reconcile such "details" of creation with belief in a benevolent first cause.[7]

By the end of the book, Darwin's refrain that "these are strange relations on the view of each species having been independently created, but are intelligible if all species first existed as varieties" became constant. Here he is, for example, on the invasion of imported species into lands long inhabited by other forms:

As natural selection acts by competition, it adapts the inhabitants of each country only in relation to the degree of perfection of their associates; so that we need feel no surprise at the inhabitants of any one country, although on the ordinary view supposed to have been specially created and adapted for that country, being beaten and supplanted by the naturalised productions from another land.[8]

That this purely natural process caused suffering, and sometimes even extinction, made sense, Darwin implied, if viewed as a purely mechanistic, naturalistic, secondary cause to which the first cause was paying no attention.

As a naturalist trained within the theory of special creation, Darwin had clearly been puzzled by more than the suffering caused by the instincts of the cuckoo, Ichneumon wasps, and slave-making ants. He also noted how natural selection meant we need not marvel "at the astonishing waste of pollen by our fir-trees."[9] Of course, by 1859 he had deeply personal experience with the fact that fir-trees were not the only organic being that suffered high mortality among its young. Having lost two infants and ten-year-old Annie, he was trying to understand why, as Ben Franklin asked a century earlier, a benevolent, designing God would create so many "exquisite Machines," "just set it a-going," only to "suddenly dash it to pieces" to be "deposited in the dark Chambers of the Grave."[10] Whether fir trees or *Homo sapiens*, Darwin was trying to make sense of why nature, as Lord Tennyson's poem lamented, "seemed so careless of the single life." He used the example of "the astonishing waste of pollen by our fir-trees" as an example of apparent imperfection, but the "astonishing waste" of human children was closer to the hearts of both Darwin and his readers.[11]

A majority of readers would have had deep personal experience with the fact that the greatest destruction within the human species fell upon the very young. Survive past the age of five, and one had a higher chance of seeing sixty. But because only about three out of every four children made it to five safely, the average lifespan in England was forty years old. Evidence existed that the child mortality rate had fallen slightly since the eighteenth century, but in Ireland, for a decade, that rate had actually *increased* since 1840. Meanwhile some argued that such heavy destruction did *not* fall upon the young of animals, thus demonstrating God's special plan for humanity. (The physician Benjamin Waterhouse, for example, cited the uniquely high mortality among human children as evidence that "child management" could be improved and premature death prevented.) But Darwin, like Malthus, insisted that the same laws governed both animals and humans: heavy destruction of the young of *all beings* was the inevitable result of the natural laws by which population increased at a higher rate than food. As we will see, Darwin did not then argue that children should not be saved. But he did believe that, once one saw human beings as governed by the same laws as the rest of nature, what had once been mysterious could be explained.[12]

In the final paragraph of *On the Origin of Species* Darwin explicitly gave his readers permission to place God behind the entire process he was proposing when he concluded his "one long argument" as follows:

> *Thus, from the war of nature, from famine and death, the most exalted object which we are capable of conceiving, namely, the production of the higher animals, directly follows. There is grandeur in this view of life, with its several powers, having been originally breathed into a few forms or into one; and that, whilst this planet has gone cycling on according to the fixed law of gravity, from so simple a beginning endless forms most beautiful and most wonderful have been, and are being, evolved.*[13]

(In subsequent editions Darwin added "by the Creator" after "originally breathed.") Some readers recovered the God who attends the fall of every sparrow within those words. But from his private correspondence we know that Darwin ultimately refused to place a close personal God in charge of such a system. In the private "autobiography" composed for his family, he commented on the view that the tremendous amount of suffering in the world was designed by a benevolent God to inspire moral improvement and, ultimately, to save souls. "The number of men in the world is as nothing compared with that of all other sentient beings," he wrote, "and these often suffer greatly without any moral improvement."

"What advantage," Darwin wondered, "can there be in the sufferings of millions of the lower animals throughout almost endless time?" (Note that Victorians tended to view children as sentient beings who, especially as infants, could not benefit morally from pain and suffering. In that sense, they were more similar to "the lower animals" than adult human beings.) He then echoed the pattern of his argument in *Origin* (comparing his theory with the theory of special creation) by noting that the tremendous amount of suffering in the world "is quite compatible with the belief in Natural Selection."[14]

These were private thoughts, but his Cambridge professors easily located the alternative explanation of suffering in *On the Origin of Species*. They did so with dismay. Although Professor Henslow loyally defended Darwin against accusations of infidelity, he drew a direct connection between Darwin's theory and the problem of suffering. Henslow wrote that he considered "an inquiry into the origin of species about as hopeless as an inquiry into the origin of evil." Both ultimately depended on trust in God's purposeful providence and close personal governance. And he believed the fact that Darwin's work "did not allow for the interposition of the Almighty" a serious objection: "God did not set the creation going like a clock, wound-up to go by itself, but from time to time interposes and directs things as he sees fit." For Henslow, maintaining God's ability to intervene in the creation of species maintained His ability to attend to human lives.[15]

Sedgwick, who had long harbored fears that belief in evolution would bestialize man, wrote candidly to Darwin of his disappointment and disapproval after reading *Origin*: "If I did not think you a good tempered & truth loving man," Sedgwick confessed, "I should not tell you that . . . I have read your book with more pain than pleasure." The crux of the matter, for Sedgwick, was the following: "I call (in the abstract) causation the will of God: & I can prove that He acts for the good of His creatures. He also acts by laws which we can study & comprehend." Because he believed this, he had hope of heaven, "but on one condition only—that I humbly accept God's revelation of himself both in His works & in His word." (At least Sedgwick had a sense of humor, for he prefaced news of his health with the words: "And now to say a word about a son of a monkey & an old friend of yours.")[16]

Stung by Sedgwick's claim that his book demonstrated "unflinching materialism," poor inductive reasoning, and a "demoralised understanding," Darwin wrote to Lyell: "If ever I talk with him, I will tell him that I never could believe that an inquisitor could be a good man, but now I know that a man may roast another & yet have as kind & noble a heart as Sedgwick's." But for Sedgwick the stakes were too high to consider his wayward pupil's feelings too much. Sedgwick believed in a universal sense of right and wrong, meaning in affliction, and a future state. And he was certain that to uphold these beliefs, God must be discernable in the Book of Nature. "There is a moral or metaphysical part of nature as well as a physical," he wrote to Darwin. "A man who denies this is deep in the mire of folly." The study of design linked the material to the moral world, yet Darwin had "ignored this link," Sedgwick admonished, "&, if I do not mistake your meaning, you have done your best in one or two pregnant cases to break it. Were it possible (which thank God it is not) to break it, humanity in my mind, would suffer a damage that might brutalize it—& sink the human race into a lower grade of degradation than any into which it has fallen since its written records tell us of its history." Darwin thought both nature and humanity could be quite brutal whether the moral and material world were linked by proof of design or not. And in the end, he had decided that dispensing with Sedgwick's vision of the world explained that fact better.[17]

Nature Groaneth and Travaileth

Reverends Henslow and Sedgwick did not think Darwin's theory could be reconciled with belief in God's close personal providence. But other quite orthodox Christians did. Indeed, historians have demonstrated that some conservative Christians found it easier to reconcile Darwin's theory with Christianity than more liberal Christians who wished to emphasize God's mercy and goodness over his justice. The most influential attempt to reconcile Darwin's explanation of the origin of species with Christian doctrine was composed by the Harvard botanist Asa Gray. At first glance Gray's support for Darwin may seem surprising, for we met Gray earlier sharply criticizing *Vestiges* for distancing God's hand from creation. But on both theological and scientific grounds, and as a friend, Gray saw Darwin's work quite differently. He explicitly argued that Darwin's theory finally reconciled the existence of great suffering, both human and animal, with God's power, wisdom, and goodness. He was not alone. Ministers like John T. Gulick agreed that Christians should find little threatening in the creation of species via the "struggle for existence," for did Scripture not say, Gulick asked, that "the whole creation groaneth and travaileth together in pain until now"?[18]

When Gray presented Darwin's work to the American public, he placed the problem of suffering at the center of natural selection's explanatory power. Gray argued, for example, that Darwin had eliminated the "common objection" that too much evil and suffering existed to believe in a benevolent God, for he had turned apparent misery, evil, and suffering to "practical account." Darwin had explained "the seeming waste as being part and parcel of a great economical process. Without the competing multitude, no struggle for life; and without this, no natural selection and survival of the fittest, no continuous adaptation to changing surroundings, no diversification and improvement, leading from lower up to higher and nobler forms." (The phrase "survival of the fittest" was not Darwin's, but the biologist, philosopher, and sociologist Herbert Spencer's. Convinced by Alfred Russel Wallace that Spencer's phrase was "an excellent expression," Darwin added "survival of the fittest" to the fifth edition of *On the Origin of Species* in 1869.)[19]

For Gray, the greatest puzzles of Paley and Sedgwick's versions of natural theology became the "*principia* of the Darwinian." He did not say humans might identify and know God's purposes in creating through so violent a process. Indeed, he thought Darwin's vision of nature aligned well with orthodox Christianity's emphasis on humanity's inability to understand the ways of God. The natural theology of men like Paley, which insisted the world was overall a happy place, had provided fodder for atheists and materialists because the world did not, in fact, reflect human concepts of either wisdom or goodness. But Christians, in Gray's view, could not expect to square such mysterious circles. Criticisms of Darwin on the grounds he had turned nature into a nasty, misery-ridden place made no sense to Gray: Darwin had simply observed what creation was like, and tried to explain nature by appealing solely to secondary causes. In doing so, Gray insisted, Darwin's work posed no new threats to theism. As a firm believer in God's sovereignty, the mystery of divine governance, and the inability of humans to scrutinize God's ways, Gray was not bothered by natural selection, especially given Darwin's implicit placement (in those opening epigraphs and the final paragraph of *Origin*) of God as the designer of the entire process.

As we have seen, emphasizing God as the first cause of a convincing secondary cause was perfectly acceptable both theologically and scientifically. Indeed, in his review of *Vestiges*, Gray had defined natural law as *how God governed*. Just as the Christian must see the birth and death of the individual as grounded in natural law yet created via the "*direct* action" of "an intelligent creative cause," Gray argued that a correct understanding of Providence collapsed the boundary between natural law and miracle completely. Then, he quoted another reviewer of Darwin's book with approval: natural laws represented "exertions so frequent and beneficent that we come to regard them as the ordinary action of him who laid the foundation of the earth, and without whom not a sparrow falleth to the ground." What, Gray asked, did the difference between Darwin and his reviewer amount to? Darwin merely inquired into the "form of the miracle," and in doing so, Gray believed, set the doctrine of the being and providence of God on firmer ground.[20]

Because they trusted each other, Gray's interpretation of Darwin's theory inspired some of Darwin's most personal reflections on the concept of a God who attends the fall of every sparrow. As Gray worked away at convincing American readers that natural selection could be reconciled with design, and thus Christianity, Darwin confessed to him:

I own that I cannot see, as plainly as others do, & as I should wish to do, evidence of design & beneficence on all sides of us. There seems to me too much misery in the world. I cannot persuade myself that a beneficent & omnipotent God would have designedly created the Ichneumonidæ with the express intention of their feeding within the living bodies of caterpillars, or that a cat should play with mice. Not believing this, I see no necessity in the belief that the eye was expressly designed.[21]

Though he was always patient with Darwin, Gray was usually unsympathetic to criticisms like this. In his review of *Vestiges* he had written that "Human conceptions of the Deity are for ever at fault in imputing to him the errors and deficiencies which belong to our own limited faculties and dependent condition." But Gray clearly preferred Darwin's theory of evolution to *Vestiges*: descent with modification via natural selection acknowledged—and indeed, was built upon—the tremendous suffering evident in creation. And to Gray that made all the difference. He firmly believed that Darwin's vision of nature mapped better onto orthodox visions of the fallen state of both creation and humanity.[22]

Gray also believed that evidence of God's goodness, which seemed to disappear if one looked at creation and human experience too closely, could be recovered in the fact that the evolutionary process was progressive: after all, it had culminated in humankind! Gray placed this vision of Darwin's theory in concert with Christianity's vision of suffering as redemptive before the American public during the lead-up to a Civil War that soon took the lives of 750,000 soldiers. Gray believed that the Civil War demonstrated Christianity's great truth that all suffering, which must be expected in a fallen creation, was ultimately redemptive. He described God as violently overturning the old corrupt order and halting the advancing tread of slavery. Clearly, wariness regarding humanity's ability to understand why so much suffering existed on Earth did not mean one didn't fight against sin. Like many evangelical abolitionists, Gray believed humans must actively oppose injustice, while trusting that Providence was on their side. He wrote of the war to Darwin as follows: "We work hard, and persevere, and expect to come out all right, to lay the foundations of a better future, no matter if they be laid in suffering." He might as well have written: *Of course* they are laid in suffering.[23]

I UNHESITATINGLY AFFIRM MY PREFERENCE FOR THE APE

Gray's argument that Darwin's theory could be reconciled with Christianity did not convince everyone. Back in Britain, one of the first reviews of *On the Origin of Species* was penned by the Bishop of Oxford, Samuel Wilberforce. Reverend Wilberforce came out firmly against Darwin's conclusions. Though his primary concern seemed to be Darwin's one-sentence concession that he included the origin of man in his theory, Wilberforce focused much of his review on Darwin's methods. He deftly argued that Darwin relied too much on imagination, rather than evidence. Given Darwin strongly hinted human beings arose from animal ancestors, Wilberforce insisted he would "scrutinise carefully every step of the argument which has such an ending, and demur if at any point of it we are invited to substitute unlimited hypothesis for patient observation, or the spasmodic fluttering flight of fancy for the severe conclusions to which logical accuracy of reasoning has led the way."[24]

Like Asa Gray, Wilberforce had no problem admitting the existence of a struggle for existence (that, he wrote, had long been known), nor that "it tends continually to lead the strong to exterminate the weak." Indeed, he found in "this law . . . a merciful provision against the deterioration, in a world apt to deteriorate, of the works of the Creator's hands." But he had a very different explanation for why the world operated this way. Wilberforce diagnosed the "real temper" of *On the Origin of Species* in Darwin's reaction to the suffering that exists in the world. The sting of the bee causing the bee's own death, the Ichneumonidae larvae eating caterpillars alive, the astonishing wasting of pollen by our fir trees, the presence of death and famine—all seemed to Darwin, Wilberforce noted, "inconceivable on the ordinary idea of creation." But a much simpler solution for the presence of these "strange forms of imperfection and suffering amongst the works of God" was available: the fall of man as told in the Bible. "We can tell him," Wilberforce wrote, "of the strong shudder which ran through all this world when its head and ruler fell."[25]

Wilberforce agreed with Darwin that the study of nature might tell one something about why suffering existed, but he came to a very different conclusion regarding the lesson to be learned. Adam Sedgwick reminded us, Wilberforce wrote, that true science "teaches us to see the finger of God in all things animate and inanimate, and gives us an exalted conception of His attributes, placing before us the clearest proof of their reality; and so prepares, or ought to prepare, the mind for the reception of that higher illumination which brings the rebellious faculties into obedience to the Divine will."[26] He had no sympathy with Darwin's rebellion against the idea that God would permit so much suffering.

Others did sympathize with Darwin's alternative explanation of the state of the world, and they soon came to his defense. In November 1859, as he waited for reviews of *Origin* to be composed and printed, Darwin wrote to his friend John Lubbock that he cared "not for reviews, but for the opinion of men like you & Hooker & Huxley & Lyell." We have already met Hooker and Lyell. But who was Huxley?[27]

Thomas Henry Huxley was a thirty-four-year-old professor of natural history at the Royal School of Mines. Darwin need not have been anxious about Huxley's opinion. Huxley later recounted how, upon reading *On the Origin of Species*, he exclaimed: "How extremely stupid not to have thought of that!" Darwin's book, he wrote, "provided us with the working hypothesis we sought. Moreover, it did the immense service of freeing us for ever from the dilemma—Refuse to accept the creation hypothesis, and what have you to propose that can be accepted by any cautious reasoner? In 1857 I had no answer ready, and I do not think that anyone else had. A year later we reproached ourselves with dullness for being perplexed with such an inquiry."[28]

Huxley was very aware that the "creation hypothesis" and its defenders would not go quietly. Having finished *Origin*, he wrote to Darwin: "As to the curs which will bark & yelp—you must recollect that some of your friends at any rate are endowed with an amount of combativeness which . . . may stand you in good stead." He assured the reclusive Darwin that he was "sharpening up my claws & beak in readiness." Soon, Darwin referred to Huxley as "my good and kind agent for the propagation of the Gospel." It was Huxley who ensured that this "gospel" permeated both British natural history and science education as quickly as it did.[29]

Although they were united by their fight for a purely natural explanation of species, in many ways Darwin and Huxley came from different worlds. Charles was a comfortable upper-class gentleman with family wealth to support his science. Huxley was a lower-middle-class professional (i.e., he was paid to do science) in a society in which the best jobs went to more well-connected men. Darwin had to travel the world to witness enormous suffering. Huxley knew the struggle for existence firsthand, in England. As a teenager he had served as apprentice to a surgeon (at the time surgeons were low in status). He witnessed terrible sights during his rounds in East and North London, sights that, as his biographer Adrian Desmond wrote, scarred him for life: "Ten crammed to a room, babies diseased from erupting cesspits, the uncoffined dead gnawed by rats."[30]

He carried a bag of drugs as he made his rounds, but those drugs were useless for most of the suffering he encountered. People who came to him for medical aid, Huxley wrote, "were really suffering from nothing but slow starvation." He collected a reservoir of painful memories.

I have not forgotten—am not likely to forget so long as memory holds—a visit to a sick girl in a wretched garret where two or three other women, one a deformed woman, sister of my patient, were busy shirt-making. After due examination, even my small medical knowledge sufficed to show that my patient was merely in want of some better food than the bread and bad tea on which these people were living. I said so as gently as I could, and the sister turned upon me with a kind of choking passion. Pulling out of her pocket a few pence and halfpence, and holding them out, "That is all I get for six and thirty hours' work, and you talk about giving her proper food."[31]

Huxley often wondered why people did not "sally forth in mass and get a few hours' eating and drinking and plunder to their hearts' content, before the police could stop and hang a few of them." A Liverpool detective replied to his question with: "Lord bless you, sir, drink and disease leave nothing in them." The memories fired Huxley's belief that the world needed a new foundation, new leaders, and new knowledge. Meanwhile he read the calls of dissenters, including the physician Thomas Southwood Smith's *Divine Government*, which argued that social reform, based upon an accurate understanding of nature's laws, was a divine duty.[32]

Darwin had learned about the struggle for existence mainly by watching others. Huxley had fought his way to medical school and then to a post as surgeon on the HMS *Rattlesnake* (1846–1850). The ship traveled the Torres Straits between Australia and New Guinea, mapping the reefs so that British ships could quickly move from one part of a growing empire to the other. Several men on the voyage died. As ship's surgeon, Huxley confronted their deaths (as he wrote when one died of blood poisoning) with the question: "Am I not more or less guilty of this man's death from want of knowledge?" This was an explanation of suffering and mortality that placed responsibility for loss of life squarely and solely on the shoulders of science, medicine, and the hands and mind of the physician who had not known enough to prevent death.[33]

During his voyage Huxley also discovered how much joy and love a man could feel, and why it might be worth living in such a world after all. He met the love of his life, Henrietta (Nettie) Anne Heathorn, at a ball in Sydney, Australia, in July 1847. They were engaged from afar for eight years, for it took Huxley that long, having returned to London, to secure a position that would allow him to send for his fiancé, marry, and start a family. Meanwhile, Huxley put Britain's nepotism-ridden society on notice through sharp and merciless attacks on the status quo. Undermining the theory of special creation, tied since the seventeenth century to defenses of the existing social order, was an important part of this fight. Upon reading *Origin*, he found the means of harnessing nature to the side of a new society in which a man obtained rewards based on merit, rather than his class and connections. Huxley wanted the worth of a man to be determined by his ability to succeed amid free and open competition among men, rather than based on who his parents or grandparents were (even if, in the deep past, his ancestors were apes).

Six months after his own positive review of *Origin* appeared, Huxley was attending the British Association for the Advancement of Science in June 1860 when he had a chance to demonstrate how sharp his beak and claws really were. Bishop Samuel Wilberforce was on the docket to speak on Saturday. Like Huxley, Wilberforce had ripped *Vestiges* to shreds fifteen years earlier. But as we have seen, Wilberforce was not convinced that Darwin's book was any more scientific or safe. Huxley hadn't intended to go to Wilberforce's lecture. He wanted to leave Oxford early and spend time with Nettie, but he happened to run into none other than Robert Chambers in the street. Seven years earlier Huxley had written a scathing review of the tenth edition of *Vestiges*. Although he thought transmutation a "perfectly legitimate" idea if a purely naturalistic mechanism could be found, Chambers's talk of natural laws as "the will of a present Deity" clearly irritated him. He called the book "a mass of pretentious nonsense" and a "notorious work of fiction." Chambers does not seem to have begrudged the criticisms. Huxley was an ally to the cause of getting evolution a fair hearing and wresting creation from Anglican theologians, a cause which, standing on an Oxford sidewalk, Chambers promptly reprimanded Huxley for deserting. "Oh! if you take it that way, I'll come and have my share of what is going on," Huxley replied. So he joined the crowd of respectable guests (and some rowdy undergraduates in the back) in Oxford's new Museum of Natural History to hear what Wilberforce had to say.[34]

Darwin had been both impressed and annoyed by Samuel Wilberforce's review of *Origin*. "Is it not grand," he wrote to Huxley, "the way in which the Bishop asserts that all such facts are explained by ideas in God's mind?!" But he also thought the review, as he reported to Hooker, "uncommonly clever." Wilberforce had picked "out with skill all the most conjectural parts, & brings forward well all difficulties." Indeed, Darwin set to writing an entire book to deal with the issues Wilberforce raised respecting domesticated animals. Wilberforce had argued that conclusions drawn from domestic animals could not be so easily extended to nature: pigeon breeds had not crossed the species boundary. On this point Huxley agreed with him. Huxley did not agree, however, with Wilberforce's assumption that a bishop had the right, authority, or knowledge to say so one way or the other.[35]

In print, Wilberforce had obviously picked out valid difficulties. He had also, albeit briefly, taken issue with the fact that, in the final pages of *Origin*, Darwin strongly implied that he applied "descent with modification" to the origin of human beings. Such a conclusion, Wilberforce had insisted, could not be reconciled with reason, observation, or Christianity. Make no mistake, he wrote: "Man's derived supremacy over the earth; man's power of articulate speech; man's gift of reason; man's free-will and responsibility; man's fall and man's redemption; the incarnation of the Eternal Son; the indwelling of the Eternal Spirit, all are equally and utterly irreconcilable with the degrading notion of the brute origin of him who was created in the image of God, and redeemed by the Eternal Son assuming to himself his nature."[36] Before a real audience in Oxford, he seems to have gotten a bit carried away with his concern that Darwin's theory threatened Scripture's message that "ye are of more value than sparrows" and, in the course of his speech, made a joke about apes at Huxley's expense. Historians aren't sure what he said, except it must have been something about Huxley's own ancestry. For having been asked to stand and reply, Huxley concluded his stern rebuttal of Wilberforce's criticisms of Darwin's book with the following extraordinary words:

If then . . . the question is put to me would I rather have a miserable ape for a grandfather or a man highly endowed by nature and possessed of great means of influence & yet who employs these faculties & that influence for the mere purpose of introducing ridicule into a grave scientific discussion, I unhesitatingly affirm my preference for the ape.[37]

As a result, Huxley became known as the man who said he would rather be descended from an ape than a bishop. The Wilberforce-Huxley exchange soon became one of the most famous tales in the so-called "warfare between science and religion." (These tales often forgot the earnest support of Asa Gray and other Christians for Darwin's theory, and Darwin's concession that Wilberforce had raised important scientific criticisms.) In fact, the battle between Huxley and Wilberforce represented a fight over what, precisely, the boundaries between science and theology should be, and who got to draw them. Wilberforce saw himself as duty-bound to comment on Darwin's theory, and by the older rules of British science, he was quite right. But Huxley was intent on changing those rules. The tight historical ties between politics, natural history, and theology meant that to do so he had to fight on several fronts. Huxley had insisted it did not matter where a man came from: what mattered was what a man did with his life and how he behaved. The "dignity of man," Huxley implied, was to be won by seeking truth rather than error and working for the sake of a better Earth.[38]

This New School of Prophets

Huxley's home with Nettie and their growing family provided a refuge from such fights. Leslie Stephen later recounted how Huxley confessed, "It was not merely that he was surrounded by a sympathy which soothed the irritation of intellectual warfare, but such a home life gave meaning to vague maxims of conduct and deepened the sense of responsibility." The summer of his exchange with Wilberforce, Nettie was pregnant with her fourth child, who would follow three-year-old Noel, two-year old Jessie, and sixteen-month-old Marian. Huxley called the eldest boy, Noel, his "fair-haired, blue-eyed, stout little Trojan, very like his mother." He looked out on the world with "bold confident eyes and open brow," Huxley reported, "as if he were its master." His father had great hopes for Noel, and was intent on creating a world in which his children, even the girls, would have a fair chance at both happiness and success based on their individual merits rather than either connections or class. (He believed the girls could not, by nature, outcompete the boys in intellect or in the public sphere, but argued that this did not mean they should not be allowed to try.)[39]

Nettie took her husband's campaigns in stride. When engaged and fearing for Huxley's life during his time on the HMS *Rattlesnake*, she had occasionally confessed her discomfort with her fiancé's skepticism. She once wrote to Huxley of her trust "in our Great Father who has so long watched over us with such tender kindness, unworthy though I be so cared for. . . . Oh that He may still continue to guard us, to preserve us for each other and to grant us many years spent happily together." But, raised in a Moravian tradition of religious tolerance, she had an intellectual and theological framework that emphasized trying to understand her husband's skepticism rather than trying to convert him. Others might call her husband an atheist, but she knew he was devoutly religious in his own way. His was not a biblical religion (though he knew the Bible, it was said, better than many of his opponents), yet Nettie knew from observation that a man might be an unbeliever yet out-moral the most devout clergyman.[40]

As Huxley campaigned to get Anglican theologians out of science, Nettie was drawn to liberal versions of Anglicanism. She gravitated, for example, toward the sermons of Llewelyn Davies, a minister whom even Huxley, who was famous for "parson-baiting," respected as one of the few clergymen informed on scientific matters. Davis was a good example of how Anglican theology was changing, and the fact Darwin had more churchmen on his side than Huxley's "warfare" tales implied. Indeed, four months after *Origin* was published, an even more scandalous (and more widely read) book than Darwin's had appeared from within the folds of the Anglican clergy. Entitled *Essays and Reviews*, it contained seven essays on Christianity, almost all by Anglican ministers intent on reforming Anglican theology so that it would not be swept away before the advance of science. The authors insisted that God must be understood as a lawgiver, with no exceptions. In this view, Darwin's hypothesis was a welcome revelation of how God chose to govern the world. Thus, one of the contributors, Baden Powell, wrote of "Mr. Darwin's masterly volume" as bringing "about an entire revolution in opinion in favour of the grand principle of the self-evolving powers of nature."[41]

As Anglican churchmen fought over the extent to which Christianity must adjust to Darwin's new history without giving up essentials of the faith, Darwin and Huxley wrote letters, strategized, compared notes on critical reviews, and worked to get the good ones (including Asa Gray's, which Darwin thought one of the best) to a wider readership. Both men wrote from households in which the cries and laughter of children filled the air. Their letters were peppered with anxious asides about the ever-present illnesses that could steal children within days or long, weary months. On September 10, a few months after Huxley's exchange with Wilberforce, Darwin wrote to Huxley of his anxiety for his daughter Etty's health. She "slowly, very slowly, but steadily improves." He and Emma were thinking of a trip to the seaside "for her & us all."[42]

A few days later, the Huxley children came down with scarlet fever, the illness that had taken one of the Darwin babies soon after Wallace's letter arrived. Noel, the eldest of the Huxley children and Huxley's "little Trojan," died on September 15, 1860, after two days of feverish, delirious illness. The final violent attack lasted two hours, and then, as Huxley wrote to his friend Frederick Dyster, "we had a dead child in our arms."[43]

Shortly after the boy's death, Huxley sat down at his desk and wrote the following below the note that, four years earlier, had recorded his birth:

> *And the same child, our Noel, our first-born, after being for nearly four years our delight and our joy, was carried off by scarlet fever in forty-eight hours. This day week [the past week, counting from today] he and I had a great romp together. On Friday his restless head, with its bright blue eyes and tangled golden hair, tossed all day upon his pillow. On Saturday night the fifteenth, I carried him here into my study, and laid his cold still body here where I write. Here too on Sunday night came his mother and I to that holy leave-taking.*[44]

Grief-stricken, they walked together in the rain so it could cool Nettie's pounding headache. For a time, Huxley feared the grief would send her into madness. Months later, Nettie herself wrote that she wondered if she would ever be "clear minded again."[45]

On September 18, Darwin received news of the Huxleys' loss, and wrote the following:

> *I cannot resist writing, though there is nothing to be said. I know well how intolerable is the bitterness of such grief. Yet believe me, that time, & time alone, acts wonderfully. To this day, though so many years have passed away, I cannot think of one child without tears rising in my eyes; but the grief is become tenderer & I can even call up the smile of our lost darling, with something like pleasure. My wife & self deeply sympathise with M^rs. Huxley & yourself. Reflect that your poor little fellow cannot have had much suffering. God Bless you. Charles Darwin.*[46]

The "God Bless you" was a vestige of old beliefs to both men by this point, yet no secular equivalent existed for what Darwin was trying to say.

Emma and Charles saw Nettie for the first time since Noel's death when she attended Huxley's lecture at the Royal Institution. They whisked her and the three remaining Huxley children to Downe for "a change of air," where Emma and Nettie read Tennyson together. Meanwhile, both

Nettie and her husband tried to bend their hearts and minds into submission, though one spoke of God and the other of nature. Upon the death of a friend, Nettie once wrote to her sister-in-law: "Wherefore this world, its creatures? Do we all just shrivel up? I have faith that all is wisely planned, however inexorable the details and anyway I have my work to do, part of which is submission, the hardest part of all." Ultimately, Nettie found solace in the "commonplace," in her daily work and all the tasks required to keep their household running and her husband's work hours quiet and peaceful.[47]

Nettie and Huxley, like most middle-class Victorian couples, had very different kinds of work to do, as they tried to wrestle their hearts into submission. Huxley wrote to Hooker that while a new baby boy was "a great blessing" to Nettie in many ways, "for myself I hardly know yet whether it is pleasure or pain. The ground has gone from under my feet once, and I hardly know how to rest on anything again. Irrational, you will say, but nevertheless natural." They might resign themselves to taking the world "as coolly" as possible, but "that coolness," he insisted, "amounts to the red heat of properly constructed mortals." Then he wrote of all the work they had to do, to advance science, and what the fight would take, including "a good deal of unquestionable pain."[48]

Trying to get back to work, Huxley canvassed developments in science in an effort to locate the means of harnessing knowledge of nature to change the world. Four days after Noel's death he wrote to Herbert Spencer about an article Spencer was writing: "You will forgive the delay . . . when I tell you that we have lost our poor little son, our pet and hope." He tried to be philosophical: "So end many hopes and plans—sadly enough, and yet not altogether bitterly. For as the little fellow was our greatest joy so is the recollection of him an enduring consolation. It is a heavy payment, but I would buy the four years of him again at the same price." Then he explained to Spencer how bird's air cells work, to make sure that Spencer, who was trying to explain society itself via evolution, had his facts straight.[49]

Within a week of Noel's death, as he tried to find something upon which to "rest again," Huxley received a letter from one of the most liberal of Anglican clergymen, Reverend Charles Kingsley. This was the same man who permitted Darwin to add his testimony that "he has gradually learnt to see that it is just as noble a conception of the Deity to believe that He created a few original forms capable of self-development into other and needful forms, as to believe that He required a fresh act of creation to supply the voids caused by the action of His laws" to the second edition of *On the Origin of Species*. Huxley knew, then, that Kingsley was a potential ally in the fight to get Darwin's work a fair hearing. Now, Kingsley wrote to him about a matter that might seem quite separate from a campaign for evolution, but in fact went to the heart of what Huxley, at least, believed evolution meant to a long-suffering world. We do not have Kingsley's letter of condolence, but we can tell from Huxley's reply that Kingsley had urged him to take consolation in the promise of heaven.

Huxley clearly appreciated Kingsley's letter, but he rebelled from Kingsley's vision of where he must find comfort. In his reply, Huxley considered the various arguments for an afterlife. One, he noted, was the argument that "the moral government of the world," which obviously entailed great suffering, "is imperfect without a system of future rewards and punishments." Another was "that such a system is indispensable to practical morality." Both these dogmas, Huxley wrote, were "very mischievous lies." Huxley believed that rewards and punishment came to men and women *on Earth*, based upon their obedience or disobedience "to the *whole* law—physical as well as moral." He did not believe that moral obedience would atone for physical sin (a good man was not any less likely to contract a terrible disease if he was ignorant of natural laws). Cause and effect ruled the physical and moral worlds, Huxley argued, and given "the ledger of the Almighty is strictly kept . . . every one of us has the balance of his operations paid over to him at the end of every minute of his existence." Waiting for a heavenly reunion in the hereafter blinded people to the very real duties of the earthly present. "I believe," he wrote, "that the seeking for rewards and punishments out of this life leads men to a ruinous ignorance of the fact that their inevitable rewards and punishments are here." It was ignorance and willful opposition to the universal laws that governed the natural and moral world, Huxley argued, that led to suffering, and not the purposes or plans of a close personal God who held the redress in reserve.[50]

Having repudiated hope of heaven, Huxley admitted to Kingsley that when Noel died, "the great blow which fell upon me seemed to stir his convictions to their foundation, and had I lived a couple of centuries earlier I could have fancied a devil scoffing at me and them—and asking me what profit it was to have stripped myself of the hopes and consolations of the mass of mankind?" Then he gave this firm reply to that imaginary fiend: "Oh devil! truth is better than much profit. I have searched over the grounds of my belief, and if wife and child and name and fame were all to be lost to me one after the other as the penalty, still I will not lie." He gestured to the one thing he found of value in Christianity, namely, the call to set aside one's own wishes and desires in order to know the truth. "Science," he wrote, "seems to teach in the highest and strongest manner the great truth which is embodied in the Christian conception of entire surrender to the will of God. Sit down before fact as a little child . . . follow humbly wherever and to whatever abysses nature leads, or you shall learn nothing."

Huxley described a deep sense of submission to Noel's loss, but not because he believed in a close personal God who directed all for some unknown purpose. "I cannot see one shadow or tittle of evidence," he wrote, "that the great unknown underlying the phenomenon of the universe stands to us in the relation of a Father [who] loves us and cares for us as Christianity asserts." Then, he tried to explain what he did believe: "If in the supreme moment when I looked down into my boy's grave, my sorrow was full of submission and without bitterness," it was "not because I have ever cared whether my poor personality shall remain distinct for ever from the All from whence it came and whither it goes." Rather, submission arose from his belief that the world was governed by natural laws. And if that was true, there was no one and nothing to blame but his own ignorance (though this was hard enough), and nothing to do but get back to work. This was why Huxley could write in his diary, as he recorded Noel's death below the lines that had recorded his birth: "I say heartily and without bitterness—Amen, so let it be."

A call to action lurked in Huxley's insistence to Kingsley, in a letter triggered by Noel's death, that "Life cannot exist without a certain conformity to the surrounding universe." He warned Kingsley that the "great and powerful instrument for good and evil, the Church of England," must acknowledge the universality of nature's laws if it was not to shiver into fragments before the advancing tide of science. Speaking of those who studied the natural world based on that universality, Huxley wrote: "Understand that this new school of the prophets is the only one that can work miracles, the only one that can constantly appeal to nature for evidence that it is right, and you will comprehend that it is of no use to try to barricade us with shovel hats and aprons, or to talk about our doctrines being 'shocking.'" ("Shovel hats" referred to a brimmed hat and aprons to a traditional vest worn by Anglican clergy.)

As historians of science have shown, Huxley was actually embracing assumptions that had governed Christian interpretations of science in Britain for generations. A commitment to the uniformity of nature's laws, and belief in progress via a better understanding of those laws, was, for many British Christians, perfectly reconcilable with their faith in God and the direction of history. (In the heat of the battle, Huxley tended to overstate the degree to which his own "side" relied upon natural laws and the other "side" on miracles.) But definite points of disagreement did exist in how Huxley, as opposed to Anglican proponents of science, explained the world. And Huxley's letter to Kingsley shows how, as Robert Chambers wrote nearly two decades earlier, those differences came to the fore at the "deathbed of amiable children."[51]

Huxley's repudiation of Christ's demand to fear not amid affliction, for not a sparrow shall fall on the ground "without your Father," was tightly tied to his campaign against the other part of Christ's demand: "ye are of more value than many sparrows." The Book of Matthew eloquently reminded Christians of the close relationship between faith in God's close personal control *and* human uniqueness. For Huxley, arguing against the assertion that God attended the fall of every sparrow, and that his governance was thus close and personal, went hand in hand with a repudiation of the belief that man was created independently of all other animals. After all, in arguing against Darwin's explanation of suffering,

Bishop Wilberforce had made the link between belief in human uniqueness and a close personal God clear:

> *Man's derived supremacy over the earth; man's power of articulate*
> *speech; man's gift of reason; man's free-will and responsibility; man's*
> *fall and man's redemption; the incarnation of the Eternal Son; the*
> *indwelling of the Eternal Spirit, all are equally and utterly irrec-*
> *oncilable with the degrading notion of the brute origin of him who*
> *was created in the image of God, and redeemed by the Eternal Son*
> *assuming to himself his nature.*[52]

In agreement with Wilberforce that Christianity's explanation of
suffering depended upon both man's creation "in the image of God" and
"man's fall and man's redemption," Huxley was the first of Darwin's allies
to place anatomical evidence that human beings came from apes before
the British public. (Darwin himself had included just one sentence hint-
ing as much in *Origin*, when he wrote that the acceptance of his views
would place psychology "on a new foundation, that of the necessary
acquirement of each mental power and capacity by gradation" and "light
will be thrown on the origin of man and his history.") In explicitly breaking the barrier between "man and animal," Huxley took on the most powerful naturalist in Britain: Professor Richard Owen. Father of William
Owen (the man who wrote to Darwin that, in the wake of his son's death,
he took comfort in the God who attends the fall of every sparrow), Professor Owen claimed that the human brain contained a unique organ, the
Hippocampus minor, possessed by no ape. Huxley boldly and publicly told
him he had his facts wrong and then published the results of anatomical dissections that located the supposedly uniquely human part in the
brains of apes. Huxley's anatomical arguments against Professor Owen
and his reply to Reverend Kingsley might seem like different campaigns.
But both arose from Huxley's commitment to a world in which natural
law, completely divested of God's attention or intervention, governed the
world, from human origins to the loss of a beloved child. There must be,
he insisted, *no* exceptions.[53]

In the years after Noel's death, Huxley cast about for a new word to describe his position on the universe. He decided on the term *agnostic* in 1869. "It came into my head as suggestively antithetic to the 'gnostic' of Church history, who professed to know so much about the very things of which I was ignorant. . . . To my great satisfaction the term took." During the last quarter of the nineteenth century, Huxley helped transform unbelief from a radical position to a respectable one for the middle classes by claiming the moral high ground in debates with believers. Agnostic "simply means," he explained, "that a man shall not say he knows or believes that which he has no scientific grounds for professing to know or believe." He declared the opposite doctrine, that some things should be taken on faith, not only wrong but immoral. (Huxley had as little patience with atheists as he had with theists; atheists, too, he argued, claimed to know things that they could not prove.)[54]

Huxley had reservations about natural selection (prior to experimental proof, he insisted, it was a good working hypothesis only), but he fought tooth and nail for Darwin's emphasis on a purely naturalistic explanation of species, including *Homo sapiens*. He wasn't just fighting for Darwin's theory. He was fighting for a young generation of professionals out to wrest science—and its paid positions—from clergymen and the upper class. He was out to get science in schools and train civil servants for an empire that, he argued, must be based purely on scientific and technological knowledge if Britain was to compete with other European powers. He was out to convince working men (via a series of "Lay Sermons") that science was "creating a firm and living faith in the existence of immutable moral and physical laws, perfect obedience to which is the highest possible aim of an intelligent being." And he was fighting for Darwin's explanation of child loss.[55]

Huxley was also fighting for the only way he believed child loss might be prevented in future: the discovery of nature's laws via the disciplined methods of experimental thinking. There was not much to show for this stance in the 1860s. Smallpox was still the only disease for which a vaccine existed. Child mortality rates seemed to be moving downward in some communities, but improvements one year were often lost the next as late as the 1890s. Hints of the importance of sanitation had been present for a while: John Snow's famous argument that cholera spread via water appeared in 1849. But after a disease entered a home, neither medical science nor physicians could do very much. Huxley and Kingsley agreed that improved knowledge of nature's laws might be able to change that terrible fact. But whether Christian or agnostic, anyone hopeful that science might save children in future relied on faith, rather than evidence, that science might "work miracles" *after* a child fell ill with scarlet fever, measles, smallpox, diphtheria, tuberculosis, or whooping cough. With one important exception (examined in the final chapter) this would not change until the next century. In the meantime, confidence that strict naturalism was a better way to understand and improve the world came from elsewhere.[56]

CHAPTER SEVEN

The Descent of Man

DARWIN HAD PLANNED TO INCLUDE A CHAPTER ON "MAN" IN HIS SPE-
cies book until 1857. He removed it after losing access to a correspondent
who had been his best source of data on human variation. A naturalist
named Edward Blyth had been sending answers to a range of queries
from India, but the Indian Rebellion, which inspired the British to take
formal administrative control over India, halted the post. Darwin's loss
of access to Blyth mattered because Darwin had developed a particular
explanation for the origin of human variation called sexual selection, and
he wanted to have as much data as possible. Without Blyth's facts, he
was stuck.[1]

After he published *On the Origin of Species* two years later, Darwin
clearly hoped that someone else would write a book applying his theory
to human beings. Driven by his commitment to consilience and certain
that evolution made sense of his own kind as well, he knew it had to be
done eventually. But his friends disappointed him one after the other.
Those for whom he had the highest hopes—Charles Lyell and Alfred
Russel Wallace—after some promising starts, refused to apply evolution
by natural selection to the human mind. Both halted at the realm of
human intellect and moral sense, which they argued required a supernatural creative force.

Darwin's reply to those who would make human beings an exception,
The Descent of Man, appeared twelve years after *Origin*, in 1871. Meanwhile, in 1869 Huxley coined the term *agnostic*, and Darwin embraced
the position. Having described himself as certain of a first cause when he

wrote *Origin*, in subsequent years "disbelief," he later wrote, "crept over me at a very slow rate, but was at last complete. . . . The mystery of the beginning of all things is insoluble by us, and I for one must be content to remain an Agnostic." The tone and language of *The Descent of Man* reflected Darwin's shifting views. It was twice as long, peppered with footnotes to well-known authorities, and confident, even defiant, in its assertion of humanity's animal ancestry. Apart from a brief section on the origin of belief in God and a few nods to the important role of monotheism in the evolution of morality, there were no references to God at all.[2]

Although she supported him in all his endeavors, Emma Darwin clearly disagreed with the purely natural account of the origin of humanity's intellectual abilities and moral sense that Darwin proposed in *The Descent*. The relationship between this disagreement and her faith in God's close personal governance can be seen in her confession to her friend Frances Power Cobbe (who would review *The Descent* for the *Theological Review*) that, "speaking in my private capacity," she thought "the course of all modern thought is 'desolating' as removing God further off." We have seen how, in 1861, Emma wrote to Charles that she found the only relief amid suffering was to take it as "from God's hand, & to try to believe that all suffering & illness is meant to help us to exalt our minds & to look forward with hope to a future state." Darwin insisted, by contrast, that only an entirely animal ancestry made sense of the world, from his own ailments to why a fever could enter a house and kill children in a few days.[3] Undoubtedly, this strict emphasis on natural laws in *The Descent* had the capacity, for better or worse (Darwin thought for better; Emma thought for worse), to divest individual children's deaths of meaning by "removing God further off." Yet ironically, the book inadvertently conferred meaning on *differential* child mortality (the fact that the children of some communities died at higher rates) in a way that allayed the potential costs of "removing God farther off" for some, and, for others, drove the costs of an evolutionary interpretation of human origins far too high.[4]

MUCH INGENIOUS SPECULATION

A few years after *Origin* appeared, Darwin confessed to his friend, the botanist Joseph Dalton Hooker, that he regretted having "truckled to public opinion" by using the "Pentateuchal term of creation, by which I really meant 'appeared' by some wholly unknown process" (*Pentateuchal* refers to the first five books of the Old Testament). The man to whom he confessed this regret was also the first friend Darwin told about natural selection, years earlier. Hooker had spoken in defense of Darwin during Huxley's famous confrontation with Bishop Wilberforce in Oxford, and he was the first naturalist to apply natural selection to the geographical distribution of plants.[5]

Hooker had also traveled the world. Like Darwin, he wanted purely naturalistic explanations of the phenomena he observed, and he wanted questions about God and immortality bracketed outside the realm of science. He told Darwin that he considered discussions of "a wise Providence" ordering nature "bosh and unscientific." "I think and believe that all reasoning upon the subject is utterly futile," he added, and "that there is no such thing in a scientific sense." However, he did add that, if we must "have a Providence in the affair . . . yours is the God one," and the Providence of the proponents of the theory of special creation was "the Devil's." In other words, Hooker thought that if any theory of species was consonant with the idea of a governing Providence, it was Darwin's, rather than Paley's, Henslow's, or Sedgwick's. Hooker wanted God distanced from each sparrow's fall, too.[6]

By the time Hooker wrote lines noting his preference for a universe governed by natural laws, he and Darwin were also tied by the tragic bond of child loss. On September 28, 1863, Hooker's second daughter, Maria (Minnie) Elizabeth, died. She was six years old. Hooker wrote to Darwin within the hour: "I think of you more in my grief, than of any other friend. Some obstruction of the bowels carried her off after a few hours' alarming illness—with all the symptoms of strangulated Hernia." "Strangely enough," Hooker wrote, "I never knew she was dying till 3 minutes before the breath left her body. For 3 hours I was blind to every one of those symptoms of rapidly approaching dissolution that every nurse knows & every novelist describes—& I have seen myself so often."

Hooker was shocked when the doctor said she was dying. Within three minutes she was gone. "The retrospect of that last night," he wrote, "is thus in some respects comforting in others hideous, & I can still feel the cold shudder that every misinterpreted symptom still sent through me, during that long night of agony & suspense." His wife, he wrote, was still weak with sorrow.[7]

Darwin replied to Hooker's black-margined notes from Malvern (where he was taking the water cure, noting that he was "very weak & can write little"): "I understand well your words: 'wherever I go, she is there.'—I am so deeply glad that she did not suffer so much, as I feared was inevitable. This was to us with poor Annie the one great comfort.— Trust to me that time will do wonders, & without causing forgetfulness of your darling." He signed off: "God Bless you my best of friends." They were soon back to writing about geology, gravel pits, botany, and natural history, but Hooker wrote a month later that "it will be long before I get over this craving for my child; or the bitterness of that last night. To nurse grief I hold is a deadly sin, but I shall never cease to wish my child back in my arms, as long as I live."[8]

A year later, as the anniversary of Mary's death approached, Hooker wrote that "as the day draws nearer I feel all the misery of last year crawling over me, and my lost child's face and voice accompany me everywhere by day and by night; so that I now dread an attack of what were more the horrors of delirium tremens than the chastened sorrows of a sensible man." He felt bad writing "on such selfish subjects" to Darwin, but explained that "your affection for your children has been a great example to me, and there is no other living soul with whom I can talk of the subject; it would make my wife ill if I went on so to her. She is wonderfully different from me, the loss simply made her very ill, almost dangerously so." He told Darwin that, still, in the main he had been right: "Time, as you told me it would, has done its inevitable work." And there was something else that reconciled him to his loss, "in a real though irrational manner." He had learned of another man's "far more dreadful blow" (probably the musician Sir George Grove, who had recently lost his ten-year-old daughter to scarlet fever). The knowledge turned his thoughts to the grief of another, and thus dampened the acuteness of his own grief.[9]

Three years later, when his beloved father-in-law died, Hooker reflected on how different the death of the young versus the old seemed. How different grief felt in the two cases! He grieved for his father-in-law, but at least loss at so advanced an age seemed somehow natural. "How different in my child's case!" Hooker wrote, "I cannot see that it is altogether natural, though it is so in the main."[10] Natural, in that sickness was governed purely by natural law; yet *not* natural in the sense that his child had not lived the lifespan that an individual might and thus should live. Why, then, did they sometimes die so soon? When explaining why death exists, William Paley had had little trouble addressing the death of a man who had lived to an old age: "Immortality upon this earth is out of the question. Without death there could be no generation, no sexes, no parental relation, i. e., as things are constituted, no animal happiness." Each animal had a "natural age" to which it might live, but all must, in the end, die.[11] Why, then, did children die before living to humanity's appointed "natural age"? Paley's reply was that, if only the old died, the age of death would be certain and life would be taken for granted, resulting in a whole train of evils. But that answer was clearly unsatisfactory to some.

When, in 1863, Charles Lyell published one of the first book-length attempts to assess the implications of Darwin's theory for the origin of humans (*The Geological Evidences for the Antiquity of Man*), he noted the puzzle raised by high child mortality. Having committed himself to supporting Darwin's work in public, he cited the fact that "one-fourth of the human race die in early infancy" to counter those who questioned evolution because it did not map onto their vision of what the world should be like. Lyell reminded anyone troubled by the "struggle for existence" that the "constitution of the world" had presented "analogous enigmas" long before Darwin wrote his book. A casual glance at the Registrar General of births and deaths in Great Britain proved "that millions perish on the earth in every century, in the first few hours of their existence." Indeed, "to assign to such individuals their appropriate psychological place in the creation," Lyell wrote, "is one of the unprofitable themes on which theologians and metaphysicians have expended much ingenious speculation."[12]

Lyell mentioned child mortality in order to remind his readers of the "painful realities of the present," realities that surely meant that theories about the *past* should not be tested "by their agreement or want of agreement with some ideal of a perfect universe which those who are opposed to his opinions may have pictured to themselves." The present did not hold up to such ideals; so why, he asked, should the past? Darwin, after all, had invented neither the struggle for existence nor the tremendous amount of animal and human suffering it entailed. He had simply tried to explain phenomena that everyone must acknowledge exists.

And yet, even as he defended Darwin's explanation of the origin of species, Lyell could not, as he joked in private, "go the whole orang" and derive humans entirely from apes. He argued that Darwin's hypothesis did *not* oblige anyone to assume "an absolutely insensible passage from the highest intelligence of the inferior animals to the improvable reason of Man." Instead, one might posit a special, creative, divine act to explain the transition from the "unprogressive intelligence of the inferior animals" to the "first and lowest form of improvable reason manifested by Man."[13]

Darwin was not impressed. He wrote to Asa Gray of his disappointment with Lyell's "timidity," and wrote to Lyell himself that the words about a creative break between man and "inferior animals" had "made him groan." Any one would think "that you were far from believing that man was descended from any animal." Lyell replied that he had gone as far as reason would allow: "But you ought to be satisfied, as I shall bring hundreds towards you, who if I treated the matter more dogmatically would have rebelled. I have spoken out to the utmost extent of my tether, so far as my reason goes, and farther than my imagination and sentiment can follow, which I suppose has caused occasional incongruities."[14]

For Darwin, insisting upon a sharp spiritual division between man and animal resulted in far more than just "occasional incongruities." The stance that men were somehow "of more value than sparrows," linked as it was to William Owen's belief that the world was governed by an "Almighty Spirit who we are bound to believe cannot err, & without whose knowledge & will not a Sparrow falls," did not make sense of the world. Writing to Hooker of Lyell's book, Darwin confessed: "I wish to Heaven he had said not a word on the subject."[15] Disappointed, Darwin shifted his hopes to the other naturalist who, unsatisfied with Paley's world, had searched for an alternative answer to the question, "Why do some die and some live?": Alfred Russel Wallace.

The Facts Beat Me

Initially, at least, Darwin had good reason to hope Wallace was the perfect man for the job. After receiving a copy of *On the Origin of Species* in 1860, while still collecting at Ternate (in present-day Indonesia), Wallace wrote to his brother George: "Mr Darwin has given the world *a new science*, and his name should, in my opinion, stand above that of every philosopher of ancient or modern times. The force of admiration can no further go!!!" Far from disappointed that Darwin had published a book first, Wallace wrote to his friend and fellow naturalist Henry Walter Bates that no matter how long he had worked at the subject, he *"could never have approached the completeness"* of Darwin's book, nor "its vast accumulation of evidence, its overwhelming argument, and its admirable tone and spirit." He really felt thankful, he insisted, "that it has *not* been left to me to give the theory to the world." And then he set to work on everything that still needed to be done, on geographical distribution, on what would soon be called "ecology," and on the origin of human beings.[16]

For a few years after Wallace returned to England, he and Darwin saw eye to eye on a great deal. They had a lot in common, for example, in their rebellions against Anglican Christianity. In writing to a friend in 1861, Wallace wrote that he had "wandered among men of many races and many religions," and "pondered much on the incomprehensible subjects of space, eternity, life and death." Early in his life, he had become an *"utter disbeliever"* in almost all that Christians held to be true. He particularly despised the idea that happiness in a future state would be a reward for belief in certain doctrines. Such a view assumed that belief is both voluntary and meritorious, but Wallace held that belief depended upon evidence, which meant it must be independent of human will or desires. To think a man would be rewarded for believing certain doctrines (or punished for disbelief) made no sense. How, he demanded, could belief be meritorious when reward and punishment were supposed to follow from so self-interested a choice? He saw much to admire in all religions. "To the mass of mankind religion of some kind is a necessity," he wrote, "but whether there be a God and whatever be His nature; whether we have an immortal soul or not, or whatever may be our state after death," surely one need not fear having to suffer in the hereafter because one searched for the truth.[17]

Fully committed to an evolutionary vision of the past, in March 1864 Wallace delivered a paper to the Anthropological Society of London that extended the theory of natural selection to the origin of man. This was in some ways a surprising venue for Wallace to choose. The Anthropological Society was the intellectual wing of polygenist defenders of American slavery (the society actually had Confederate spies among its members), and Wallace was no defender of slavery. He was, however, very interested in debates over both the origin of human variation and the boundary between humans and animals, both topics the society discussed. And he was clearly, at the time at least, no egalitarian. Standing before the "Anthropologicals," Wallace confidently applied his original question ("Why do some die and some live?") to human groups, as he tried to imagine how human beings had evolved from apes. Groups competing with groups had, he argued, led to both intellectual and moral advance over time. Just look about the world: "The same great law of *'the preservation of favoured races in the struggle for life,'*" Wallace announced, "leads to the inevitable extinction of all those low and mentally undeveloped populations with which Europeans come into contact. The red Indian in North America, and in Brazil; the Tasmanian, Australian and New Zealander in the southern hemisphere, die out, not from any one special cause, but from the inevitable effects of an unequal mental and physical struggle."[18]

An anthropologist (and Spiritualist) in the audience named Carter Blake questioned Wallace's conclusion that it was "a fact" that "nations that have been extirpated by other nations. . . . were inferior, either intellectually or physically, to the nations that came after them." Wallace replied with confidence: "It appears to me that the mere fact of one race supplanting another proves their superiority. . . . We cannot tell what causes may produce it. . . . But still there is the plain fact that two races came into contact, and that one drives out the other. This is proof that the one race is better fitted to live upon the world than the other." The same processes that explained the animal and plant world, Wallace insisted, explained the human world as well. "Just as the more favourable increase at the expense of the less favourable varieties in the animal and vegetable kingdoms," he argued, and "just as the weeds of Europe overrun North America and Australia, extinguishing native productions by the inherent vigour of their organization, and by their greater capacity for existence and multiplication," so, too, did Europeans advance around the globe.[19]

Within just a few years, Wallace would move closer to Blake's view that other factors than "intellectual or physical superiority" were at play, but in the meantime, Darwin (who we will see was not opposed to talk of European superiority) was impressed. Indeed, he offered to send Wallace his references and notes "on man" if he planned to "follow out your views" and write a book on the subject.[20]

Soon, however, Darwin learned that Wallace had changed his mind about the ability of natural selection to explain humanity's intellectual, social, and moral capacities. Wallace never abandoned Darwin's theory outside the realm of the human mind (indeed, he tended to out-Darwin Darwin in his loyalty to natural selection in the nonhuman animal realm). He always believed that natural selection had answered the great questions of taxonomy that had first turned him to the problem of the origin of species. But like Asa Gray and Charles Lyell, Wallace ultimately decided that Darwin's theory did *not* provide a consilient explanation of the human mind.[21]

Wallace adopted the stance that only a spiritual soul could explain man's rational and moral sense for two reasons: First, sitting through a series of séances in 1865, he witnessed a range of phenomena, from table-rapping to apparitions of the dead, that purely materialistic accounts of the world could not, in his view, explain. He saw an accordion play itself, furniture move, and a pen write messages with no hand attached to it. He saw flowers appear in a room while the "spiritualist medium" (Mrs. Guppy) was held down as a test. He saw the medium Mr.

Home place hot coals in the hands of scores of witnesses without harm. Such phenomena, Wallace noted, had been attested again and again: "They are facts, of the reality of which there can be no doubt; and they are altogether inexplicable by the known laws of physiology and heat." The existence of a spiritual realm provided the only explanation for what he had observed. "The facts became more and more assured," he recounted, "more and more varied, more and more removed from anything that modern science taught or modern philosophy speculated on." "The facts," Wallace firmly concluded, "beat me."[22]

Second, Wallace decided that evolution via natural selection could not explain the human mind. Darwin himself, Wallace pointed out, held that natural selection could not develop an organ that would be valuable in future generations if at present it was useless. Well, Wallace demanded, of what possible benefit to humanity's ancestors was a brain that would one day invent calculus, compose symphonies, develop complex notions of justice and abstract reasoning, and imagine God? And yet (and now he spoke *against* the fundamental assumptions of the "Anthropologicals"), the brains of even the "lowest savages," with "mental requirements" analogous to many animals, were of comparable size and complexity to "the highest types," the average European. Natural selection, Wallace argued, "could only have endowed the savage with a brain a little superior to that of an ape, whereas he actually possesses one but very little inferior to that of the average members of our learned societies." In other words, *all* humans, on average, possessed extraordinary rational and moral capacities. And that meant, according to Wallace, that there must be an "Overruling Intelligence" guiding the "great laws of organic development" toward some "unknown, noble destiny."[23]

In 1869 Darwin expressed frank dismay that Wallace had altered his views. The fact his friend and co-discoverer, to whom he had once offered his notes on man, had abandoned a purely naturalistic explanation of the human mind was confounding. Darwin had a hard time not taking it as a personal betrayal. Facing the prospect of reading Wallace's latest article, Darwin let the frustration slip: "I hope you have not murdered too completely your own & my child." After reading it, Darwin wrote: "As you expected I differ grievously from you, and I am sorry for it. I can see no necessity for calling in an additional and proximate cause in regard to Man." A year later, Darwin still could not temper his disappointment: "But I groan over Man—you write like a metamorphosed (in retrograde direction) naturalist, and you the author of the best paper that ever appeared in the *Anthropological review*! Eheu! Eheu! Eheu!—Your miserable friend, C. Darwin."[24] (The exclamation means "Alas! Alas! Alas!")

Huxley might have stepped in to counter the heresies of Lyell and Wallace, of course, but he had enough work to do at the School of Mines, two Royal Commissions (including the Royal Commission on Contagious Diseases), and the London School Board. Meanwhile, he was defending the idea of man's ancestry from apes by focusing on similarities of anatomy and physiology. As a man who had to work for a living while also trying to completely reorganize British science, he simply did not have time to attack the final citadel of the claim "ye are of more value than sparrows": the origin of human intellectual and moral capacities.[25]

Darwin now knew he would have to write the book on Man himself.

No Right to Expect Immunity

The Descent of Man, or Selection in Relation to Sex was published in February 1871 and sold out in just a few weeks. Darwin did not address the evolution of humanity's rational and moral sense right away. Instead, he began with several chapters aimed at collapsing any presumed anatomical boundaries between humans and the "lower animals." He surveyed the facts of comparative anatomy, homology, and rudimentary organs, and repeatedly asked the question he had posed in *Origin*: Which explanation—descent with modification, or special creation—made better sense of the facts?

In *The Descent*'s opening, Darwin wrote that there was one naturalist whose work, had he known of it before, would have rendered his own book unnecessary: the German naturalist Ernst Haeckel's, whose book *General Morphology* had appeared in 1866. "Almost all the conclusions at which I have arrived," Darwin wrote, "I find confirmed by this naturalist, whose knowledge on many points is much fuller than mine." Darwin particularly relied on Haeckel in his discussion of rudimentary (or vestigial) organs. In describing the vermiform appendage (the appendix), Darwin wrote: "Not only is it useless" in man, as opposed to other mammals, "but it is sometimes the cause of death, of which fact I have lately heard two instances: this is due to small hard bodies, such as seeds, entering the passage, and causing inflammation." Haeckel, he added, had "remarked on the singular fact of this rudiment sometimes causing death." (That "singular fact" had first been noticed two years after Darwin left Edinburgh by the Quaker physician Thomas Hodgkin, who had composed one of the first clear descriptions of acute appendicitis in an account of an autopsy of a student who had died of enteritis after experiencing severe pain on the lower right side of the abdomen.)[26]

This was a puzzle indeed, for how could an organ for which no function could be found, but which occasionally, after hours of agony, killed its possessor, be explained? Darwin offered the following reply:

In order to understand the existence of rudimentary organs, we have only to suppose that a former progenitor possessed the parts in question in a perfect state, and that under changed habits of life they became greatly reduced, either from simple disuse, or through the natural selection of those individuals which were least encumbered with a superfluous part, aided by the other means previously indicated.[27]

Darwin generally allowed his readers to draw their own conclusions regarding any potential theological implications of such phenomena. Ernst Haeckel, by contrast, told his readers exactly what they must conclude. In *The Evolution of Man*, first published in 1876, Haeckel wrote of how such rudimentary organs "strikingly refute the teleology of certain philosophers." (By *teleology* Haeckel meant declarations that purposeful parts demonstrated God's design.) *Dysteleology* (the study of the absence of purpose via comparative anatomy), Haeckel argued, was "of the utmost philosophical importance." "In the intestines," he explained, "we have a process that is not only quite useless, but may be very harmful—the vermiform appendage. This small intestinal appendage is often the cause of a fatal illness." Such organs, Haeckel concluded, "completely shattered . . . the ancient legend of the direct creation of man according to a pre-conceived plan and the empty phrases about 'design' in the organism."[28]

In private correspondence Darwin and Haeckel found even more common ground. A few weeks after his beloved wife Anna Sethe died (her symptoms indicate the culprit may have been appendicitis), Darwin wrote Haeckel to thank him for a copy of his masterpiece, a monograph on the radiolaria, adding that he hoped Haeckel could "work hard on science & thus banish, as far as may be possible, painful remembrances." (How Darwin learned of Anna's death is not recorded.) Six months later, Haeckel wrote that he had found in On the Origin of Species "all at once the harmonious solution of all the fundamental problems that I had continually tried to solve ever since I had come to know nature as she really is." A stroke of fate having "destroyed all prospects of happiness in my life," Haeckel declared himself ready to dedicate his life to "one goal . . . namely to disseminate, to support and to perfect your theory of descent." This was a long-standing tradition among the men in Darwin's family, at least: wrestle grief into submission by actively bolstering an alternative explanation of the world. Darwin stated his explanation most eloquently in the midst of arguing that humans are, like animals, subject to the struggle for existence. He acknowledged that it was impossible not to regret the existence of famine, disease, death, and war that resulted from this inevitable struggle: "But as man suffers from the same physical evils as the lower animals, he has no right to expect an immunity from the evils consequent on the struggle for existence." He was made of the same stuff, and subject to the same physiological laws. And that was the only reason why children, and young wives, died.[29]

Encouraged by the support of Haeckel, Huxley, and others (and sustained by Emma's care and careful proofreading), Darwin confidently moved from the vermiform appendix to the human mind. Each mental trait supposed to separate man from animals—curiosity, imitation, attention, memory, imagination, reason, progressive improvement, use of tools, abstraction, self-consciousness, language, sense of Beauty, superstition—could also be found, Darwin insisted, in nonhuman animals. Dogs dreamed (demonstrating they had imagination); horses found their way home (they possessed memory); monkeys felt grief (they felt attachment to others). "So intense is the grief of female monkeys for the loss of their young," he wrote, "that it invariably caused the death of certain kinds kept under confinement by Brehm in North Africa." Even precursors to man's "moral sense" (the ability to form ideas about right and wrong) and conscience (the ability to judge one's conduct and feel remorse), could be found in nonhuman social animals. Did not dogs seem capable of learning how they should and should not behave?[30]

Now, having collapsed any presumed physical and mental boundaries between humans and animals, Darwin had to explain *how* the appearance of so great a gulf (and he conceded the gulf seemed wide) between humans and nonhuman animals had arisen. He had opened *The Descent* with a list of what must be established in order to bring human beings within the explanatory power of his theory: Did man vary "in mental faculties"? Were those variations transmitted to offspring according to the same laws as in animals? Could man be classified, "like so many other animals," into different varieties and races? Did man tend "to increase at so rapid a rate, as to lead to occasional severe struggles for existence; and consequently, to beneficial variations, whether in body or mind, being preserved, and injurious ones eliminated"? And finally, "Do the races or species of men, whichever term may be applied, encroach on and replace one another, so that some finally become extinct"?[31]

As we have seen, Darwin had traveled the world as a committed abolitionist who tended to see non-Europeans as more like him than not. When he wrote *The Descent* he still viewed monogenism (that is, a single origin for all human beings) as the most appropriate taxonomic stance to take on his own kind. He also expressed himself as very skeptical of the groupings some naturalists (and politicians) used to organize human beings. Against the polygenists (who argued the genus *Homo* could be divided into separate species), Darwin pointed out that naturalists had studied man more carefully than any other organic being, yet they could not agree on how many groups might exist. Some naturalists said there were five races; some said there were sixty-three (and everything in between). "This diversity of judgment does not prove that the races ought not to be ranked as species," Darwin offered, "but it shews that they graduate into each other, and that it is hardly possible to discover clear distinctive character between them." Any good, cautious naturalist knew the correct conclusion to draw: "He will end by uniting all the forms which graduate into each other as a single species; for he will say to himself that he has no right to give names to objects which he cannot define." In other words, those who declared races different species, or proclaimed with confidence that there were three, five, or sixty-three human races, were quite simply very poor taxonomists.[32]

As he did so often in *The Descent*, however, what Darwin gave with one hand (human unity) he took away with the other (by emphasizing racial divergence). Darwin's argument for how, precisely, human intellect and moral capacity had advanced meant he needed to describe natural groups; place those groups on a hierarchy of intellectual and mental ability; and have some groups outcompete other groups amid a pervasive, violent struggle for existence. After all, one of Darwin's arguments in *On the Origin of Species* was that, if one could observe functional gradations in any particular trait in the present, one could imagine how even the most complex, purposeful trait might have evolved through "a long series of gradations in complexity, each good for its possessor." Darwin now used the same argument to convince his readers that, as he wrote in the summary of his account of love, memory, attention, curiosity, reason, and other traits in animals, the "difference in mind between man and the higher animals, great as it is, certainly is one of degree not of kind." Classifying human groups along a ladder of intellectual and moral advance allowed him to portray "a long series of gradations in complexity" in human mental and moral development. (Recall that there was one trait that Darwin did not place on a gradation: the ability to feel pain. That trait he conferred on all human beings equally, without qualification.) Thus, Darwin pushed some human communities closer to animals, constantly writing of "higher races" and "lower races" (with "higher" generally meaning those most like him in values, appearance, and behavior). As a result, he ended up with an effectively polygenist ladder upon which he placed the races of a single, "unified" species: *Homo sapiens*. Indeed, Darwin thought that "when the principle of evolution is generally accepted, as it surely will be before long, the dispute between the monogenists and the polygenists will die a silent and unobserved death."[33]

When combined with his long-standing commitment to mono-genism, Darwin's strategy meant he drew upon his memories of Jemmy Button and his limited experience with men of African descent in highly ambiguous ways. Sometimes Darwin spoke of how closely the Fuegians were to Englishmen. "The Fuegians rank amongst the lowest barbar-ians," he wrote, "but I was continually struck with surprise how closely the three natives on board H. M. S. Beagle, who had lived some years in England, and could talk a little English, resembled us in disposition and in most of our mental faculties." Later in the book, he repeated this pattern of highlighting both difference and resemblance in the same sen-tence. "The American aborigines, Negroes and Europeans are as different from each other in mind as any three races that can be named," he wrote, "yet I was incessantly struck, whilst living with the Fuegians on board the 'Beagle,' with the many little traits of character, shewing how similar their minds were to ours; and so it was with a full-blooded negro with whom I happened once to be intimate." These statements fended off polygen-ist claims that no conclusions about human evolution could be drawn from non-European races because they were separate species. But at the same time, Darwin ranking both Jemmy and John Edmonstone's "races" "amongst the lowest barbarians" allowed him to imagine how a transition from ape to "savagery" to "civilized" might have happened.[34]

So, for example, he repeated the story, told by "the old navigator Byron," of a Fuegian who had "dashed his child on the rocks for dropping a basket of sea-urchins." The difference in "moral disposition," Darwin wrote, between such a "barbarian" and Britain's most famous abolitionists, and "in intellect, between a savage who uses hardly any abstract terms, and a Newton or Shakespeare," was great, and "differences of this kind between the highest men of the highest races and the lowest savages, are connected by the finest gradations. Therefore it is possible that they might pass and be developed into each other." (Less than a decade earlier the fugitive abolitionist William Craft had pointed out, during a meeting of the British Association for the Advancement of Science no less, that, though he had lived in England for several years, he "had met very few Shakespeares." Darwin made no mention of this piercing criticism of the common "savage" versus "Shakespeare" trope.)[35]

At least one reader of *The Descent* pointed out that it was not necessary to go to the other side of the earth to find "savages" who beat their children or wives: "A visit to the Police-courts of one of the first cities in the civilised world," wrote Charles William Grant, "would show us savages who throw their wives under bridges into a river, who thrust them under drays, who knock them down and fracture their skulls, with as little remorse and for as slight provocation, as dropping a basket of sea-urchins." In other words, British men, "supposed to be of the same race" as the most virtuous abolitionist, could be "barbaric" too. Grant pointed out that to make a claim that savage minds differed so much from civilized minds, *by nature*, one would need to raise a savage amid civilization, and a child of a so-called "civilized" race amid savagery and then, once they reached the age of thirty or forty, see how far their intellectual and moral sense differed. Without such an experiment, one could say nothing about the inherent capacities of either, or place one "naturally" lower than the other.[36]

Darwin replied to many critics in the second edition of *The Descent of Man*. He did not reply to Grant's concern.

THE EXTRAORDINARY MORTALITY OF THE YOUNG CHILDREN

During the voyage of the HMS *Beagle*, Darwin had been quite willing to place his own group quite low on the "ladder of civilization." Repelled by both slavery and active campaigns to exterminate non-Europeans, Darwin had sometimes denoted Europeans the true "savages." Indeed, his moral rebellion against a world in which some human beings moved around the globe and enslaved or exterminated others had arguably inspired Darwin's original search for an alternative to both the theory of special creation and the fatherly God to which that theory had been linked. A good, wise, just God surely did not explain such a world. After noting that he did not think in another half-century there would be a "wild Indian" left north of the Rio Negro, he had written of how "it is melancholy to trace how the Indians have given way before the Spanish invaders." The governor of the town of St. Fe's favorite occupation, he recorded with disgust, was hunting Indians: "A short time since, he slaughtered forty-eight, and sold the children at the rate of three or four pounds apiece." How could men commit such atrocities? As the voyage continued, he reflected on how "It was melancholy at New Zealand to hear the fine energetic natives saying, that they knew the land was doomed to pass from their children. Every one has heard of the inexplicable reduction of the population in the beautiful and healthy island of Tahiti since the date of Captain Cook's voyages." Observing how the original inhabitants of Australia seemed to welcome the advance of the white man because they could borrow dogs from the farm houses and barter for milk from the cows, Darwin thought them "blinded by these trifling advantages." Meanwhile the white man "seems predestined to inherit the country of his children."[37]

Now, forty years after being puzzled by such things, Darwin largely set aside the moral outrage that had originally inspired his questions and transformed into the seemingly distant, scientific describer of his own kind. He had his answer to why such things happened: the inevitable, constant struggle for existence that placed human groups in conflict with each other, even as social instincts solidified self-sacrifice and cooperation *within* groups. Indeed, by the 1870s, Darwin needed human groups to be in competition with each other for his entire theory of the evolution of the moral sense to work: individuals competing with each other amid a struggle for existence could not explain the evolution of selflessness (selfless individuals, Darwin presumed, would fail in competition with selfish ones). But a moral sense *could* arise from groups competing with other groups, according to Darwin, as follows:

> *When two tribes of primeval man, living in the same country, came into competition, if (other circumstances being equal) the one tribe included a great number of courageous, sympathetic and faithful members, who were always ready to warn each other of danger, to aid and defend each other, this tribe would succeed better and conquer the other. . . . A tribe rich in the above qualities would spread and be victorious over other tribes: but in the course of time it would, judging from all past history, be in its turn overcome by some other tribe still more highly endowed. Thus the social and moral qualities would tend slowly to advance and be diffused throughout the world.*[38]

Darwin's explanation of the rise of "social and moral qualities" depended upon a number of observations, including the tendency of individuals to apply moral rules (such as "thou shall not kill") to their own group but not to those outside their group. To make a uniformitarian argument, he also needed to be able to describe groups (or what, as a naturalist, he called *varieties* and *races*) in the present as in conflict with one another, and he needed to demonstrate differential mortality between groups. In other words, unable to go to the deep past to prove that human "varieties" had "encroached and replaced" one another, Darwin looked at the world in which he lived. And what he saw in the present was some "varieties" seemingly out-reproducing other "varieties." He noted, for example, that between 1858 and 1872, the New Zealand Maori had decreased by 32.39 percent. Forty years was all it took in the Hawaiian Islands (Darwin called these lands the Sandwich Islands) for the population to decrease "sixty-eight percent!" by 1872. "When Tasmania was first colonized," Darwin wrote, "the natives were roughly estimated by some at 7000 and by others at 20,000. . . . In 1834 they consisted of forty-seven adult males, forty-eight adult females, and sixteen children, or in all of 111 souls." In 1847 the British moved the surviving thirty-six adults and ten children to Oyster Cove in the southern part of Tasmania, but Darwin reported that the "change of site did no good. Disease and death still pursued them." By 1864, one man and three elderly women alone survived. "The infertility of the women is even a more remarkable fact than the liability of all to ill-health and death," Darwin wrote, for "at the time when only nine women were left at Oyster Cove, they told Mr. Bonwick, that only two had ever borne children: and these two had together produced only three children!" (The exclamation marks are Darwin's.)[39]

Trying to make sense of what he called "this extraordinary state of things," Darwin ran through various possible explanations of population decline. Some blamed the "profligacy of the women" (probably a reference to sex for pleasure combined with the use of birth control and abortion), native wars, "the severe labour imposed on conquered tribes," and newly introduced diseases. Others, he noted, blamed the "depression of spirits" that had plagued groups like the Tasmanians when the British forcibly removed them from their homelands. A "Mr Fenton" attributed the Maori's decreasing numbers to "the unproductiveness of the women" and "the extraordinary mortality of the young children." Some attributed the latter to "the neglect of the children by the women," but Darwin had good reason to doubt that explanation. (His theory of the evolution of the moral sense depended upon mothers' love for their infants.) He thought more likely explanations would be found in a combination of the inability of "savages" to change their habits, the depression that set in when moved to unfamiliar lands, the "new diseases" that caused much death, and the "evil effects from spirituous liquors." But the "most potent of all the causes of extinction," he concluded, "appears in many cases to be lessened fertility and ill-health, especially amongst the children, arising from changed conditions of life." Owing to those changed conditions, children were born with an "innate weakness of constitution."

Notice that Darwin rooted both the "unproductiveness of the women" and differential child mortality in the *constitution* (i.e., the material bodies) of Indigenous people amid changed conditions. As always, having broken the animal-man boundary, he drew evidence for this conclusion by drawing analogies with the "lower animals." In trying to determine the origin of variation, for example, Darwin had written much about the tendency of changed conditions to influence the reproductive system, leading "both to beneficial and to evil results." Writing of changes in the "habits of life" of Pacific Islanders, Darwin noted: "Although these changes appear inconsiderable, I can well believe, from what is known with respect to animals, that they might suffice to lessen the fertility of the natives." For unlike "the civilized races," he wrote, "Man in his wild condition seems to be in this respect almost as susceptible as his nearest allies, the anthropoid apes, which have never yet survived long, when removed from their native country." Among "the wilder races of man," changed conditions or habits led to decreased health, and "in several cases the children are particularly liable to suffer." "Savages of any race," he added, who were "induced suddenly to change their habits of life" became more or less sterile, "and their offspring suffer in health, in the same manner and from the same cause, as do the elephant and hunting-leopard in India, many monkeys in America, and a host of animals of all kinds, on removal from their natural conditions." It was all, Darwin concluded, "the same problem" as to why non-native species replaced various aboriginal species. To bolster his case, Darwin noted that he was not the only one to draw this analogy: "The New Zealander seems conscious of this parallelism, for he compares his future fate with that of the native rat now almost exterminated by the European rat." "Decreasing numbers," Darwin concluded, "will sooner or later lead to extinction; the end, in most cases, being promptly determined by the inroads of conquering tribes."

In New Zealand, Australia, and Tasmania, the "conquering tribe" was, of course, the British. Darwin thus drew on the patterns of imperial and colonialist expansion as evidence of how "extinction follows chiefly from competition of tribe with tribe, and race with race." The severity of the struggle for existence, he argued, explained why, when two tribes met, the "weaker tribe" would decrease until it became extinct, via either absorption into the more successful tribe or slaughter. Thus, in searching for a purely natural explanation for the differential mortality rates between British colonizers and the Indigenous peoples of Australia, New Zealand, and Tasmania, Darwin attributed the outcomes entirely to the natural laws governing physiology, reproduction, and disease amid an intense but inevitable law-bound struggle for existence.

Nearly four decades earlier, during the voyage of the *Beagle*, Darwin had found the different mortality rates of Europeans and non-Europeans in places like Tasmania and New Zealand both "melancholy" and, ultimately, unworthy of a good God. But in the 1870s, confident of his ability to group and rank human beings, Darwin believed he had a convincing alternative explanation of why such things happened. A few hints exist in *The Descent* that Darwin still lamented the results. In acknowledging the curiosity that "savages did not formerly waste away" before the "classical nations, as they now do before modern civilized nations," he noted that otherwise "the old moralists would have mused over the event." In noting that no such "musings" existed, Darwin clearly acknowledged the presence of a profound moral problem. But notice that he was also delegating explicit, public discussions of what *ought* to happen to "moralists," and portraying his own work as purely a matter of natural history.[40]

And yet Darwin was moralizing about the world in *The Descent of Man*. He had decided that a struggle for existence provided a better explanation of suffering than the belief that an all-wise, all-powerful, and all-good God of Christianity would allow such things as part of His creation. But while he conceded that "moralists" might lament the extinction of "savages," he ultimately held that overall the process that led to such extinction was beneficent because, in his view, it led to human progress. Some kinds of suffering, Darwin implied, must be borne, because they pressed the human species upward. In doing so, as historians Adrian Desmond and James Moore note, Darwin was assuming "an inevitability that had to be explained, not a socially sanctioned expansion that had to be questioned."[41]

Darwin did not see himself as writing a commentary in defense of colonialism, imperialism, or even capitalism. Rather, he considered himself a scientific naturalist describing the world as the natural, inevitable outcome of purely natural processes (rather than the purposeful result of a designing God). But ultimately, *The Descent of Man* naturalized the belief that, overall, Europeans were "more advanced" and thus justified in overwhelming everyone else. We know from Darwin's private correspondence that he had no qualms about writing of race extermination as a means of progress. When, in 1862, Charles Kingsley wrote to him of "inferior" races dying out "by simple natural selection, before the superior white race," Darwin questioned neither Kingsley's assumptions nor his language. "It is very true what you say about the higher races of men, when high enough, replacing & clearing off the lower races," Darwin replied. "In 500 years how the Anglo-saxon race will have spread & exterminated whole nations; & in consequence how much the Human race, when viewed as a unit, will have risen in rank."[42]

Darwin clearly relied on pervasive racist assumptions about the world to make sense of the differential mortality of those whose lands the British occupied. (Indeed, this kind of thinking was so pervasive that it did not even have a word to distinguish it from other kinds of thinking. The word *racialism* was coined in 1880, *racialist* in 1902, and *racist* in 1919.) Of course, Darwin did not originate the assumptions upon which racism depends. One did not need evolution to place either individuals or "races" into conflict or on a hierarchy. After all, in the US South, slavery had been defended for generations by citing its presence in the Bible. As we have seen, those who wanted more "scientific" justifications for ignoring the suffering and mortality of some had cited Malthus's "struggle for existence" since at least the 1820s, no evolution required. But in an age of increasing admiration for science and purely natural explanations, Darwin provided useful imagery for colonizing powers intent on justifying expansion around the globe. Some, like Kingsley, placed God behind this "progressive," natural-law-governed process. Others, like Ernst Haeckel, did not. But in grouping human beings and then placing those groups on a ladder of supposed moral advance, Darwin naturalized the suffering caused by the policies of British and other imperial powers.

He did the same close to home. When considering whether natural selection still operated within "civilised countries" like Britain, for example, he expressed agreement that "the reckless, degraded, and often vicious members of society, tend to increase at a quicker rate than the provident and generally virtuous members." Then, he repeated the words of William Greg that "the careless, squalid, unaspiring Irishman multiplies like rabbits," while "the frugal, foreseeing, self-respecting, ambitious Scot" married late and had few children and, as a result, "In the eternal 'struggle for existence,' it would be the inferior and LESS favoured race that had prevailed—and prevailed by virtue not of its good qualities but of its faults."[43]

An unknown Irishman took Darwin to task for repeating Greg's words. He called the passage unworthy of Darwin, for it allowed Greg "to do for you, what in no instance, as far as I am aware, have you done for yourself—viz generalize from insufficient data." Darwin left the passage in. He did, however, think Greg wrong about one thing. In contrast to Greg's prediction that "the inferior and LESS favoured race" would prevail, Darwin insisted that "some checks to this downward tendency" existed. "The poorest classes crowd into towns," he wrote, where, as the Annual Report of Births and Deaths in Scotland showed, "during the first five years of life," the death rate in towns was double that of rural areas.[44]

Darwin did not explicitly condone this high child mortality rate among the poor in towns. As we have seen, he clearly lamented the great evils that resulted from the struggle for existence. But even as he conceded it was impossible not to "bitterly regret" the suffering that resulted from the struggle for existence, he pointed out that, had man not "been subjected during primeval times to natural selection, assuredly he would never have attained to his present rank." Thus, even as he divested creation of God's attention to the fall of every sparrow, he implied that readers might still find a benevolence in the world, so long as they looked at nature from the right perspective. Statements like this, that the struggle for existence had ultimately resulted in human progress, ultimately implied that so high a "town death-rate" might, like the high mortality rates of Indigenous communities, in the end be a good thing.[45]

In private correspondence Darwin's approval of both struggle and race extinction became even more explicit over the years. Writing to William Graham in 1881 to express his appreciation for Graham's *The Creed of Science*, Darwin wrote: "I could show fight on natural selection having done and doing more for the progress of civilisation than you seem inclined to admit." As an example, he wrote that the "more civilised so-called Caucasian races have beaten the Turkish hollow in the struggle for existence." He then offered a prediction: "Looking to the world at no very distant date, what an endless number of the lower races will have been eliminated by the higher civilised races throughout the world." Darwin criticized Graham on one point only, namely the idea that "the existence of so-called natural laws implies purpose. I cannot see this." Darwin saw natural selection (when applied to human beings) as in the main a good, albeit completely blind, force. After all, it had somehow led to European civilization! He refused, however, to put a benevolent God behind either the process or the result.[46]

Of course, Darwin himself had witnessed the active extermination campaigns carried out amid colonialist expansion by men like General Rosas in Argentina. He clearly thought such things should not happen in a civilized society. But his moral rebellion against active extermination did not translate into a rebellion against what he had decided was an inevitable, natural process of the "most fit" groups outcompeting the "less fit." Instead, he described the long process of group competition and extinction as the ultimate source of "the most noble of all the attributes of man," the moral sense or conscience. With a few important exceptions (slavery and active extermination campaigns), Darwin did not imagine these processes might or should be changed by human agency. He may have concluded that "the most potent of all the causes of extinction appears in many cases to be lessened fertility and ill-health, especially amongst the children, arising from changed conditions of life," but that one might *stop* changing conditions of life—say, by calling a halt to British colonial and imperial expansion—was an unimaginable proposition to both Darwin and many of his British readers.

In the end, Darwin was wrong that those he designated "lower races" would eventually become extinct due to their physical "constitutions." In 1924, four decades after Darwin's death, the Maori physician Te Rangi Hiroa pointed out that, just as the native fern of Aotearoa did not give way to the European clover, the Maori did not, in fact, give way to British colonizers. "The present generation," Te Rangi Hiroa wrote, "refuses to comply with the picturesque but illogical simile of following the way of the vanishing Maori rat and the extinct Maori dog." Darwin was right that extraordinary declines had occurred (the Maori population fell by over 50 percent between Cook's arrival in 1769 and 1858). But he was not correct in his fatalistic conclusion that high mortality rates arose from some inevitable, "constitutional" decline in fertility that would exterminate entire nations before the advance of the "Anglo-Saxon race." Those who did supposedly disappear (the Tasmanians and Fuegians) were actively exterminated by men who called upon God, evolution, or sometimes a powerful mix of both to explain and justify the deaths of Indigenous children and their parents.[47]

Ironically, Darwin himself had offered an explanation of "groupishness" (or what we would now call *racism, ethnocentrism, xenophobia,* and *nationalism*) that might have undermined its power. Recall that in his earliest notebooks, Darwin had reflected upon the tendency of human beings to place not only their own kind, but their own group, on a pedestal above others. "Has not the white Man," Darwin wrote in 1837, "who has debased his Nature & violates every best instinctive feeling by making slave of his fellow black, often wished to consider him as other animal?" Darwin's notebook jottings display at times a radical humanitarianism that, in combination with his awareness of man's tendency to dehumanize those outside his own group, might have prevented him from lowering *anyone* on the ladder (or even imagining a ladder). Although the indignation evident in Darwin's 1830s notebooks and *Journal of Researches* is harder to find in *The Descent*, the following attempt to make sense of human groupishness does appear:

> *As man advances in civilisation, and small tribes are united into larger communities, the simplest reason would tell each individual that he ought to extend his social instincts and sympathies to all the members of the same nation, though personally unknown to him. This point being once reached, there is only an artificial barrier to prevent his sympathies extending to the men of all nations and races. If, indeed, such men are separated from him by great differences in appearance or habits, experience unfortunately shews us how long it is, before we look at them as our fellow-creatures.*[48]

Darwin clearly held that groupishness was a social instinct that had once been critical to the evolution of humanity's moral sense. But in describing this groupishness as a social instinct that could be conquered by "the simplest reason," Darwin's words contained the radical potential to undermine that instinct. Applied to white Victorians' tendency to group and rank human beings, those words might have inspired more Victorian readers to doubt both designations of "constitutional differences" between human beings and declarations that some children's mortality was outside the boundaries of concern. This passage was not, however, the one that many of Darwin's readers remembered.

Chapter Eight

Dealing with Descent

According to an American geologist who visited Darwin in 1871, Darwin was "much impressed by the general assent with which his views had been received." Twenty years ago, Darwin reported, "such ideas would have been thought subversive of everything good, and their author might have been hooted at, but now not only the press, but, from what my friends who go into society say, everybody, is talking about it without being shocked."[1]

A large part of *The Descent of Man*'s power and influence arose from the fact that, much like the Bible, it could be drawn upon for diametrically opposed campaigns, something that, as historian Diane Paul notes, "Darwin's own ambiguities, hesitations, and waverings" made easy. He was often ambiguous, for example, regarding whether placement on some "ladder of civilization" was due to nature (in which case a group's status as higher or lower could not be changed) or nurture (in which case it was malleable and subject to education). When he spoke of "Man" having advanced so far, sometimes he seemed to be thinking of *Homo sapiens* as a whole and at other times only of Anglo-Saxons. One passage might emphasize cooperation as the source of all moral progress, and another competition. Sometimes he seemed to be simply describing the world. Sometimes he implied that nature contained clear messages about what the world, and society, should be like. Indeed, historian Richard Hofstadter notes that "if there were, in Darwin's writings, texts for rugged individualists and ruthless imperialists, those who stood for social solidarity and fraternity could, however, match them text for text with some to spare." In other words, *The Descent* could become (almost) whatever its readers wished it to be. Therein lay its power, possibilities, and dangers once scientists, physicians, and theologians began applying Darwin's vision of human evolution to the problem of child mortality and loss.[2]

Darwin's naturalization of capitalist, colonialist, and imperialist assumptions and policies helps to explain *The Descent*'s extraordinary appeal to many Victorian readers, despite the potential threats to belief in the God who personally attended to the fall of every sparrow. On the other hand, it also meant that some, though they might have found Darwin's alternative explanation of individual child loss comforting, faced the prospect that those in power might declare their own children not worth saving in the name of progress.

Scientific Moonshine

When copies of Darwin's book first arrived in the United States a year and a half prior to the outbreak of the Civil War, Frederick Douglass was in the midst of a lecture tour in Great Britain. He had extended his stay because he was under suspicion for having helped organize John Brown's raid on Harper's Ferry and the governor of Virginia had issued an order for his arrest. He took the risk of returning home when one of his daughters, ten-year-old Annie, "the light and life of my house," succumbed to a disease of the brain that had "baffled the doctors."[3]

By all accounts, though Douglass's grief when his daughter died was great, he did not find solace in belief in heaven. Sixteen years after Annie's death, when his daughter Rosetta's six-year-old daughter Allie died, Douglass urged Rosetta not to suffer from "the superstitious terrors with which priest craft has surrounded the great and universal fact of death." Death, he wrote, "is the common lot of all—and the strongest of us will soon be called away. It is well! Death is a friend, not an enemy." He would not, he explained, write of heaven: "Whatever else it may be, it is nothing that our taking thought of it can alter or improve. The best any of us can do is trust in the eternal powers which brought us into existence, and this I do, for myself and all."[4]

We do not have a record of Douglass's response to *Origin*. Like many Americans, he had other things on his mind in 1860, and other battles to fight in the years immediately following the Civil War. But we do know that he was not certain, given the society in which he and his children lived, that embracing an animal ancestry was the best means of securing justice for all.

Douglass had plenty of reasons (better reasons than even Darwin) to be skeptical of both Christianity and the theory of special creation. Both had been used to uphold a suffering-ridden status quo. Douglass often recalled how, as an enslaved child, he was told that "'God up in the sky' had made all things, and had made black people to be slaves and white people to be masters. I was told that God was good, and that He knew what was best for everybody." Given his experience of the world, the last seven words, he wrote, "went against all my notions of goodness." Later, on the abolitionist lecture circuit, Douglass delivered a lecture entitled "Slaveholder's Sermon," in which he mimicked the "Christian" slave-holder's command: "Oh, consider the wonderful goodness of God! Look at your hand, horny hands, your strong muscular frames, and see how mercifully he has adapted you to the duties you are to fulfill . . . while to your masters, who have slender frames and long delicate fingers, he has given brilliant intellects, that they may do the *thinking*, while you do the *working*." Douglass clearly had no interest in supporting such arguments. He was drawn to explanations of the world that emphasized governance via natural law and "the possibility of attaining happiness here on earth by obedience to natural laws" rather than those that emphasized God's close personal providence and waiting for heaven.[5]

It might seem that Douglass would readily adopt Darwin's theory of evolution as useful in the fight against those who cited God's designing hand to defend slavery. But by the time Darwin published *Origin*, Douglass knew that science could be a fickle friend of emancipation when practiced in a society willing to move some human beings outside the realm of the Golden Rule. For years he had watched one of his best friends, Dr. James McCune Smith, fight against those who insisted that differential mortality rates between children of European and African descent arose from their "constitutional differences." (Smith was the first US man of "mixed race" to obtain a medical degree, although he had to go to Edinburgh to study medicine, since no US medical school would admit him on account of his ancestry.) As early as 1838, Smith was taking on white physicians who insisted children of African descent did not flourish in the United States due to their "peculiarity of constitution," rather than, as Smith pointed out, the fact they grew up in environments ridden with "hell-born prejudice." The questions to be asked when establishing the cause of child mortality did not include what is the child's race, Smith argued, but where did the child live, how were they treated, and how did society view them?[6]

Smith's arguments were largely ignored by white physicians, naturalists, and politicians, who persisted in appealing to differential mortality rates as evidence of "constitutional" (i.e., material, physiological, and anatomical) differences between races. In 1854 the Alabama physician Josiah Nott published, with George Gliddon, a book entitled *The Types of Mankind*, in which he used differential mortality rates between children of European, African, and mixed descent to claim an impenetrable barrier between human races. Data on "hybrid" children, Nott claimed, showed they died young. Unlike James McCune Smith, Nott did not blame the prejudice-ridden, violent environments into which these children were born. Instead, he blamed "the fact" their parents belonged to different *species* (lower fertility was supposed to constitute proof of species boundaries). By implication, no one could do anything about these children's deaths, except to prevent "miscegenation" in the first place. Nott clearly hoped medicine (for some children) could be improved via better knowledge of natural laws. He wrote *Types of Mankind* a year after a yellow fever epidemic took four of his own children in five days. Searching for the cause of such tragic losses, he was the first physician to suggest a link between yellow fever and mosquitoes. But in rooting differential mortality rates in "constitutional differences," Nott justified action to prevent the death of some children and inaction in response to that of others, on the grounds that the latter's deaths were the inevitable result of natural laws.[7]

Most of Nott's arguments against all races belonging to one species were aimed at biblical defenses of monogenism, since most abolitionists defended human unity on scriptural grounds. But well aware that some were citing evolutionary books like *Vestiges of the Natural History of Creation* to "scientifically" unite all races via a common ancestor, Nott covered all his bases: whether of one origin or of many, the present status of the races must be, for all practical purposes, separate. "The Caucasian, Ethiopian, Mongol, Malay and American may have been distinct creations, or may be mere varieties of the same species produced by external causes acting through many thousand years: but this I do believe, *that at the present day the Anglo-Saxon and the Negro races are, according to the common acceptation of the terms distinct species, and that the offspring of the two is a hybrid.*" In other words, Nott insisted that natural history, anatomy, and physiology all justified pushing two million Americans and their children permanently outside the boundaries of the Golden Rule.[8]

With so much at stake, Frederick Douglass had weighed in on the debate over human origins in 1854. Douglass knew who would be pressed closer to animals by the American powers-that-be if the boundary between Man and Animal was collapsed. In a speech entitled "The Claims of the Negro, Ethnologically Considered," he argued that, by any measurement or trait, men of all races shared the uniquely human characteristics of any other "race." "Away with all the scientific moonshine," he concluded, "that would connect men with monkeys; that would have the world believe that humanity, instead of resting on its own characteristic pedestal—gloriously independent—is a sort of sliding scale, making one extreme brother to the ou-rang-ou-tang, and the other to angels, and all the rest intermediates!"[9]

Six years later, Charles Darwin's *On the Origin of Species by Means of Natural Selection, or the Preservation of Favoured Races in the Struggle for Life* arrived from England. As Douglass grieved his daughter Annie, the nation hovered on the brink of civil war (the war began eighteen months later). One of the primary points of contention as Americans digested Darwin's theory entailed the implications for debates over the origin and meaning of human variation. Clearly, the reasons for Douglass to withhold assent to evolutionary visions of nature, despite his belief that nature's laws, not God, governed the fall of every sparrow, had only increased since he had called connecting men with monkeys "scientific moonshine." For although some radical abolitionists cited *Origin* as proof of their belief in human unity, others used Darwin's work to emphasize divergence, whether humans shared a common ancestor or not. After the war, critics and defenders of slavery alike cited Darwin's theory as evidence that, after emancipation removed artificial barriers to labor competition through emancipation, the races would be in natural competition with each other. (As we have seen, for a generation slavery's defenders had cited Malthus's "struggle for existence" to make similar claims.)[10]

Take Thomas Henry Huxley, whose essays were widely read in the United States. Although Huxley defended abolitionism on the grounds that in a moral society "all classes of mankind shall be relieved from restrictions imposed by the artifice of man, and not by the necessities of nature," he clearly believed that emancipation had leveled the playing field enough for "natural laws" to now sort men according to their natural capacities. In 1865, as the Civil War came to an end, Huxley argued that now that the field had been leveled between the "Black and the White," nature could take its course. He was also absolutely certain that white men would, in the end, win the day in open competition with anyone else, explicitly dismissing the arguments of some "on the winning side" that on average men of all races were equal. "It is simply incredible," he wrote, "that, when all his disabilities are removed, and our prognathous relative has a fair field and no favour, as well as no oppressor, he will be able to compete successfully." By arguing that open competition would sort human beings by ability, Huxley harnessed Darwin to a version of economic liberalism that came with both blinders and guardrails useful to those in power. He assumed that all it took to "level the playing field" was to abolish slavery. The war had decided the issue, Huxley argued. All responsibility for the result of freedom must now lie between nature and those who were now free: "The white man may wash his hands of it, and the Caucasian conscience be void of reproach for evermore."[11]

As an agnostic, Huxley never put a benevolent God behind these arrangements, but others did. To do so they drew upon a long tradition of describing secondary causes as designed by the first cause, God. Historians Desmond and Moore note, for example, that Charles Kingsley believed "the 'lowly' races were Providentially doomed and that the whites would sweep out all before them to usher in God's Kingdom." Such thinking was pervasive among those in power in the United States.

In 1867 a US congressional investigation of "Indian depopulation" concluded that the causes of decline were disease, intemperance, war, loss of land, and "the irrepressible conflict between a superior and inferior race." As one expert witness testified during the hearings: "The causes which the Almighty originates, when in their appointed time He wills that one race of men—as in races of lower animals—shall disappear off the face of the earth and give place to another race . . . has reasons too deep to be fathomed by us. The races of the mammoths and the mastodons, and the great sloths, came and passed away: the red man of America is passing away!" The statement was an extraordinary mix of belief in the benevolence of natural laws; a (convenient) faith that God had reasons "too deep to be fathomed"; and firm certainty that, given any (presumably misguided) ameliorative effort would be up against both Nature and God, nothing could or should be done to prevent such suffering.[12]

Whether men cited God as having designed the world this way or not, such moves placed Frederick Douglass in a difficult position. He clearly agreed with Darwin that a world governed by natural laws made better sense of human experience than belief in the all-powerful, all-knowing, and omnibenevolent God of orthodox Christianity. Indeed, one of Douglass's heroes (of whom he kept a bust on his mantel) was the radical German theologian David Strauss (one of the most influential proponents of Higher Criticism). In 1873, Strauss captured why Darwin's work mattered so much to those trying to change the world (which, clearly, Douglass was trying to do). Philosophers and critical theologians had declared the extermination of miracles over and over again, Strauss wrote, but always the argument that special creation—a miracle—was required for the purposeful parts of animals and plants undermined those declarations. But Darwin had "opened the door by which a happier coming race will cast out miracles, never to return." Strauss argued that appeal to God's personal governance, supposedly demonstrated by the theory of special creation, dampened efforts to ameliorate suffering via political and social reform. He contrasted belief in the fall of man and human depravity with the vision offered by Darwin's theory of the origin of human beings: rather than being the descendants of a couple created in the image of God who had been kicked out of Paradise, humanity could now be proud of having gradually worked itself up "by the continuous effort of innumerable generations from miserable, animal beginnings to its present status." For Strauss, the primary difference between these historical narratives lay in what they allowed humanity to imagine of the future: "Nothing dampens courage as much as the certainty that something we have trifled away can never be entirely regained," Strauss wrote. "But nothing raises courage as much as facing a path, of which we do not know how far and how high it will lead us yet."[13]

Douglass firmly agreed with Strauss that an emphasis on the uniformity of natural laws was useful to reformers because it destroyed the old story of the fall of man and replaced it with a vision of his moral and material ascent. In a lecture delivered in 1883 on the "philosophy of reform," Douglass argued that "So far as the laws of the universe have been discovered and understood, they seem to teach that the mission of man's improvement and perfection has been wholly committed to man himself. . . . It does not appear from the operation of these laws, nor from any trustworthy data, that divine power is ever exerted to remove any evil from the world, how great so ever it may be." The law, in other words, in all directions was imperative and inexorable. If God occasionally intervened via miracles, then men could never learn nature's ways, nor adjust accordingly in the interest of either progress or amelioration. Work, planning, and activity, and trusting in a consistent system of cause and effect, were the only means provided by Providence to fight against evil and injustice and the only means of securing a better Earth.[14]

Douglass often criticized belief in individual salvation on the grounds it directed attention away from the possibility of ameliorating earthly suffering. But he also knew that men might use new ideas about both God and nature to justify radical change for some while solidifying the status quo for others. Indeed, having embraced the view of nature as a place of struggle, competition, and progress, sometimes the very abolitionists who had fought at Douglass's side to end slavery thought their work done after the war. Like Huxley, many white abolitionists believed that nature's laws would now have free reign to sort out the fit and unfit in a struggle for existence. Some, including statistician Jesse Chickering, had argued for decades that prejudice, not nature, decided who lived and who died on American soil. "The free coloured population of the United States," Chickering wrote in 1853, "seem doomed to perish in a state of freedom, falsely so called, among a people whose feelings and whose prejudices loathe fellowship with them, and where the whole structure of society is set against their equal participation in the blessings of the land." In other words, the field had not been leveled at all. But with important exceptions (including radical socialists, African-American Christians, and some white evangelical Protestant and Catholic Christians), many Americans ignored arguments like Chickering's. Meanwhile, talk of salvation via science turned evolution into a new justification of social hierarchies by assuming those hierarchies arose from the natural, benevolent result of the struggle for existence (some placed God behind the struggle, some did not).[15]

Douglass had too much experience with how science could be called upon to establish boundaries around people, naturalize hierarchies, and justify differential mortality rates as inevitable to wholeheartedly accept evolutionists' attempts to give a purely naturalistic explanation of the origins of man. He knew what happened when nature's laws, or what naturalists believed to be nature's laws, were applied to human variation within a hierarchal society ridden by bigotry and prejudice. And so, in his lecture on the philosophy of reform, Douglass expressed tremendous ambivalence when he briefly referred to Darwin's vision of the past. "I do not know that I am an evolutionist," he announced. He did confess having far "more patience with those who trace mankind upward from a low condition, even from the lower animals, than with those that start with him at a high point of perfection and conduct him to a level with the brutes. I have no sympathy with a theory that starts man in heaven and stops him in hell." But in contrast to Strauss, Darwin, and Huxley, Douglass was not willing to publicly embrace evolution. After all, he had little reason to believe that theories of "descent with modification via natural selection," as used by his contemporaries, provided the best means of saving all children.[16]

Navigating a world in which many Americans cited nature, Scripture, or some powerful mix of both to justify ignoring suffering, Douglass knew that an assessment of Darwin's work must include the question: Who would benefit from evolutionary visions of progress and who might be left behind? Within societies organized on racial, ethnic, and class-based lines, talk of progress via natural law could inspire commitment to learning about natural laws in order to ameliorate suffering, while relegating some human beings to a "lower stage" and then explaining their suffering, and even extinction, as the inevitable, "beneficent" outcome of those same natural laws. In 1884 the abolitionist lawyer Albion Winegar Tourgée lamented that many Northerners believed "the negro would disappear beneath the glare of civilization" and that "such disappearance would be a very simple and easy solution of a troublesome question." Indeed, the belief in the extinction of the Black "race," notes historian John S. Haller, "became one of the most pervasive ideas in American medicine and anthropological thought during the late nineteenth century." In other words, evolutionary visions of the Book of Nature could be used (much like the Book of Scripture) to change the world for some children, while ignoring the suffering of others. Whether one believed a choice must be made between the two books, or that they could be combined, individuals like Douglass clearly had to weigh a different set of factors than those immediately willing to "sharpen their beaks" in Darwin's defense.[17]

Is It a Devil That Has Made the World?

The same year *The Descent of Man* left the presses, a young woman in a loveless marriage "alone, fought with Death for my child." Both of Annie Besant's children had come down with whooping cough. Also known as chin-cough or *tussis convulsiva* (in modern vaccination schedules, it's called pertussis and is the "P" in the DTP vaccine), the "whoop" referred to the sound that accompanied the child's cough as they struggled to bring air into their lungs. The disease was also particularly traumatic to witness. During the worst convulsions, wrote one physician, "the patient appears to gasp, and strain, and hiccup, and strangle rather than cough. In this painful action, the veins of the face, as well as the eyes appear swelled; the countenance becomes very dark; the eyes are turned upwards, and there seems danger of a fatal strangulation." Sometimes, breath simply could not be drawn, "in consequence, of which, the patients are, as it were, suffocated."[18]

Two-year-old Arthur came through the illness without much trouble, but the attending doctor did not expect little Mabel, who had been delicate from birth, to live. For Besant, plagued by an unhappy marriage and the first hints of religious doubt, Mabel's illness became her Rubicon. The doctor had told her that recovery was impossible, and that in one of the terrible paroxysms of coughing, Mabel must die. "For weeks she lay in hourly peril of death," Besant later wrote: "We arranged a screen round the fire like a tent, and kept it full of steam to ease the panting breath; and there I sat, day and night, all through those weary weeks, the tortured baby on my knees." The most terrible part was that even a drop or two of milk brought on the "terrible convulsive choking, and it seemed cruel to add to the pain of the apparently dying child."[19]

Eventually the doctor began dabbing Mabel's face with chloroform during the worst of her convulsions. "It can't do any harm at this stage," Besant recalled him saying. "And it checks the suffering." He had not expected to see the child alive again, but Mabel slowly recovered. The effect on her mother of this experience, however, had only just begun. A century earlier, Erasmus Darwin had jeered at the idea of a close personal God who attended the fall of every sparrow by imagining that instead it must be a devil who sent whooping cough among the beautiful children of his patients. Nearly a hundred years later, Annie Besant demanded in the wake of her daughter's illness: "Is it a Devil that has made the world? . . . Is all blind chance, is all the clash of unconscious forces, or are we the sentient toys of an Almighty Power that sports with our agony, whose peals of awful mockery of laughter ring back answer to the wailings of our despair?"[20]

It took three years and two months, but eventually Besant became an atheist. During the time between Mabel's illness and her public affirmation of unbelief, Besant plunged into a severe depression, as she wrestled with the prospect of infinite emptiness. "It was the long months of suffering through which I had been passing," she later wrote, "with the seemingly purposeless torturing of my little one as a climax, that struck the first stunning blow at my belief in God as a merciful Father of men." Having lived through her "helpless, sinless babe tortured for weeks and left frail and suffering," she added other indictments, especially the suffering of the poor and the injustice of women's status in society. Besant's prior belief in God's constant direction of affairs now turned her against Him. For if she truly believed, she had to see his "individual finger" in her baby's agony, her own sufferings, and all the sufferings of the poor. But "the presence of pain and evil in a world made by a good God; the pain falling on the innocent, as on my seven months' old babe; the pain begun here reaching on into eternity unhealed" all drove her to hate rather than love God: "All the hitherto dormant and unsuspected strength of my nature rose up in rebellion; I did not yet dream of denial, but I would no longer kneel."[21]

Like so many others we have met, Besant's rebellion was not initially inspired by science; it was inspired by orthodox explanations of suffering. Her initial protest, she wrote, was "of the conscience rather than of the brain." Many Victorians saw atheism as equal to immorality. Besant insisted atheism was, in fact, *more* moral than Christianity. It was the atheist, she wrote, who was awakened "to what the world was, to the facts of human misery, to the ruthless trap of nature and of events over the human heart, making no difference between innocent and guilty." In the truly moral being, Besant argued, experience must knock belief in a close personal God out of one's heart. Clearly, the Anglican doctrine of eternal damnation disturbed her greatly, for she asked: "Given a good God, how can He have created mankind, knowing beforehand that the vast majority of those whom He created were to be tortured for ever? Given a just God, how can He punish people for being sinful, when they have inherited a sinful nature without their own choice and of necessity?" But "worst of all puzzles" was "the existence of evil and of misery. . . . It seemed so impossible to believe that a Creator could be either cruel enough to be indifferent to the misery, or weak enough to be unable to stop it. The old dilemma faced me incessantly: 'If He can prevent it and does not, He is not good; if He wishes to prevent it and cannot, He is not almighty.'"[22]

Besant declared both conservative and liberal ministers' attempts to solve these puzzles "special pleading." A minister who conceded ignorance by replying to her queries that "we have no key to the 'mystery of pain' excepting the Cross of Christ" was the only one who she respected and whose words gave her some comfort. But the same minister's advice that she take comfort in the promise of heaven (he quoted the lines from Tennyson's *In Memoriam*: "Reach a hand through time to catch / The far-off interest of tears") meant nothing to Besant. The promise of some blessed reunion in future was no longer good enough. She had lost any meaning in her own sufferings, any justification for why she should stay in a loveless marriage, and any purpose to a long-suffering world. For at least one terrible evening, the only way out seemed to be suicide.[23]

Ridicule that she, as a woman, would dare wrestle with questions that "had puzzled the greatest thinkers and still remained unsolved" did not help. "Surely it was a woman's business to attend to her husband's comforts and to see after her children," she was told, "and not to break her heart over misery here and hell hereafter." Besant was up against the Victorian middle class's famous "ideology of separate spheres," an ideology that established strict roles for wives and husbands: men must compete in the marketplace, while women must uphold morality and religion by taking care of the home and children. Breaking these rules came with serious consequences. In 1873, after she refused to take communion, her husband, an Anglican vicar named Frank Besant, expelled her from their home. (As Besant often pointed out, English marriage law gave women no rights to property, legal counsel, their own bodies, or their own children.)[24]

At first, Besant maintained custody of Mabel, "the sweetness and joy of my life" (Arthur was at boarding school), and she had access to a small income. But she had given up home, friends, and social position in favor of freedom of conscience. Having acted upon what she believed, she decided to find "something to do." She learned what the study of natural law might offer to a suffering world from the infamous atheist Charles Bradlaugh's Sunday lectures at the Hall of Science. Her conscience and heart, she wrote, had rebelled from belief in God and "the spectre of an Almighty Indifference to the pain of sentient beings." "The injustice, the cruelty, the inequality, which surround me on every side" was too much to believe in God. But she still believed in man: "In man's redeeming power; in man's remoulding energy; in man's approaching triumph, through knowledge, love, and work."[25]

Besant decided that this "approaching triumph" could occur through one means and one means only: "Study nature's laws, conform to them, work in harmony with them." As the ultimate application of natural laws to the world, evolution, for Besant, became key to making sense of the universe. Evolution explained evil as vestiges of humanity's animal past, and established the route to amelioration in the further evolution of the social instinct, growth of conscience, and strengthening of man's mental and moral nature. Antisocial tendencies and selfishness, for example, could be explained as "bestial tendencies" inherited from man's brute ancestry. All could be eliminated as men adopted "rational 'co-operation with Nature'" and the duty of mutual aid to one's fellows. Looked at through an evolutionary lens, Besant insisted, the struggle against evil "loses its bitterness" and becomes "full of hope instead of despair." For since "Evil comes from ignorance" it could be rectified via better knowledge. This was Robert Chambers without the God-talk of which Sedgwick had been so suspicious. Besant was not interested in being polite or pretending to be pious. The stakes, she decided, were too high.[26]

Besant eventually risked everything, including custody of her beloved children, to defend atheism as leading to a more moral and just society. She also explicitly enlisted Darwin's vision of nature in her campaigns, certain that particular reforms followed logically and scientifically from his new explanation for the origin of man. The most infamous and radical of those campaigns was for wider and cheaper access to information on contraception (the term *birth control* was not in use until after 1914). Publishing information on contraception in expensive medical books purchased primarily by physicians, who could thus control access to that knowledge, was legal. But Besant and her ally, the famous atheist Charles Bradlaugh, published a pamphlet on the topic (actually a reprint of an 1832 book by the American Charles Knowlton) called *The Fruit of Philosophy: The Private Companion to Married Couples* that sold for a sixpence. They were arrested and put on trial in 1877, during which a jury decided the pamphlet violated the laws against obscenity (the conviction was later reversed).

What did a campaign for contraception have to do with ideas about God and nature's laws, evolutionary or otherwise? For much of the nineteenth century, the British powers-that-be would have insisted the answer must be "very little." When, in editions from 1798 to the 1820s, Thomas Malthus wrote of the great disparity between the rapid rise of population and the slower rise of food supplies, he had not mentioned contraception as a possible means of preventing the struggle for existence. Indeed, he had categorized all artificial interference in sexual intercourse as *vice*, and believed that moral restraint (i.e., "strict chastity" before delayed marriage) was the only means of slowing population growth. "Promiscuous intercourse"—that is, sex for anything other than procreation—weakened the affections of the heart, he argued, and degraded the female character. On this point Malthus was an Englishman of his time. Few British Victorians wrote about, much less defended, contraception. One of Malthus's biographers notes that it was no more possible for Malthus to envisage birth control as a solution to population increase as it was for him "to envisage a successful flight to the moon."[27]

One can find a few exceptions to the consensus. In 1823, at the age of seventeen, the philosopher John Stuart Mill distributed information on contraception to maidservants after he had found a dead newborn baby in a park in London. (He was put in jail as punishment.) Mill insisted that surely contraception was better than giving mothers no other choice but infanticide. Indeed, Malthus had cited the practice of infanticide when responding to critics who insisted the pressure of population growth was surely not so great. Darwin too had cited infanticide among "primitive" and "barbarous" societies as evidence of the struggle for existence. John Stuart Mill, by contrast, looked at his own society, a society in which a mother might kill her child and leave it wrapped in blankets in a park, and asked who, in fact, were the real savages? Mill's question was, of course, rather inconvenient and generally ignored by those in power. After all, belief in widespread infanticide *outside* Britain provided a convenient justification of imperialism. Historian Adrian Desmond notes that the belief that Indigenous Australians murdered their young convinced Huxley that their extermination was justified. The few native Australians consulted on the question invariably insisted that infanticide was rare, but from the 1840s on British colonists spread tales of infanticide far and wide. (In doing so they rationalized the infamous abduction of children as acts of "philanthropy and rescue," a practice that continued, sanctioned by the state, well into the 1960s.)[28]

For those willing to imagine pregnancy might and *should* be prevented by other means than chastity, options existed but were generally viewed as unrespectable, immoral, and impious. *Coitus interruptus*, pessaries, condoms, and herbal potions taken orally or externally to either prevent pregnancy or induce abortion had been used around the world for a long time, with variable effectiveness and safety. When rubber was first vulcanized in 1844, condoms and cervical caps could be mass produced. Eventually, advertisements for "Malthusian devices" appeared, clearly a term Malthus would have hated. Eventually, "Malthusian" became code for anyone in favor of disseminating information on contraception to the "masses." Critics made the term synonymous with the words *immoral*, *antisocial*, and *atheist*. Besant insisted, by contrast, that contraception was the only moral and scientific means of truly helping the poor. In their preface to the pamphlet, she and Bradlaugh wrote that, given Reverend Malthus's demonstration that population increases faster than the means of existence, "*some* checks must therefore exercise control over population." "The checks now exercised," they pointed out, "are semi-starvation and preventable disease; the enormous mortality among the infants of the poor is one of the checks which now keep down the population." They were simply advocating that there were better, more scientific, more moral means of checking population. "We think it more moral," Besant and Bradlaugh insisted, "to prevent the conception of children than, after they are born, to murder them by want of food, air and clothing." Indeed, they concluded that it is "a crime to bring into the world human beings doomed to misery or to premature death."[29]

Both Besant and Bradlaugh saw their campaign for contraception as a direct outcome of Darwin's view of nature. Indeed, in preparation for their trial, Bradlaugh subpoenaed Darwin, certain that Darwin's views would support their own. In doing so, in the words of historians Desmond and Moore, he made "a gigantic miscalculation." Darwin wrote to beg that to appear as a witness in court would cause him great suffering on account of his ill health, and in any case, he would *not* be able to speak in favor of their cause. Though he had not read their pamphlet, he understood, he wrote, that "it refers to means to prevent conception. If so I sh^d be forced to express in court a very decided opinion in opposition to you, & Mrs Besant; though from all that I have heard I do not doubt that both of you are acting solely in accordance to what you believe best for mankind.—I have long held an opposite opinion . . . & this I sh^d. think it my duty to state in court." (Darwin added that he also believed "that any such practices would in time spread to unmarried women & would destroy chastity, on which the family bond depends; & the weakening of this bond would be the greatest of all possible evils to mankind.")[30] Clearly nervous that he would be called to testify, he included the following extract of *The Descent of Man* to convince Bradlaugh and Besant that he would be of no aid to their cause:

> *The enhancement of the welfare of mankind is a most intricate problem; all ought to refrain from marriage who cannot avoid abject poverty for their children, for poverty is not only a great evil, but tends to its own increase by leading to recklessness in marriage. On the other hand, as Mr. Galton has remarked, if the prudent avoid marriage, whilst the reckless marry, the inferior members tend to supplant the better members of society. Man, like every other animal, has no doubt advanced to his present high condition through a struggle for existence, consequent on his rapid multiplication, and if he is to advance still higher it is to be feared that he must remain subject to a severe struggle; otherwise he would sink into indolence, and the more gifted men would not be more successful in the battle of life than the less gifted. Hence our natural rate of increase, though leading to many and obvious evils, must not be greatly diminished by any means.*[31]

This exchange between Besant, Bradlaugh, and Darwin shows how the actions required after having dispensed with orthodoxy in favor of an evolutionary vision of nature and humanity were far from obvious. Darwin had found solace solely in the recognition that humans were no exception to the misery caused by the natural law of population. He was not against medicine, but he did think that some interventions must be weighed against the progress that arose via the struggle for existence.

(Darwin made plain in his letter that by "any means," he also meant "artificial means of preventing conception.") Besant, by contrast, countered Darwin's view by pointing out that in fact, in Europe at least, improvements in health and sanitation had resulted in population increases. Meanwhile, wages fell, unemployment was high, and food prices were increasing. "Nature" was already being interfered with, Besant argued, and a good thing, too. But given that fact, artificial, "scientific" checks such as contraception were needed to replace the (increasingly removed) "natural" checks to population growth. Like Darwin, Besant approved of medicine and sanitation, but she firmly believed that something must be done, based on reason and science, to avoid such progress from creating *more* suffering.[32]

Besant paid a high price for her convictions. In 1878 she lost custody of eight-year-old Mabel as a direct result of her campaign for contraception and defense of atheism. The court argued that Besant's unbelief made her an "unfit guardian" for her child, who would surely be "an outcast in this life and damned in the next." "The little child was carried away by main force, shrieking and struggling," Besant wrote, driven almost to madness by grief: "At last health broke down, and fever struck me, and mercifully gave me the rest of pain and delirium instead of the agony of conscious loss." Besant's rights to visit her children were eventually restored, but when she realized how those visits kept Mabel "in a continual state of longing and fretting for me," she resolved to visit no more: "Resolutely I turned my back on them, and determined that, robbed of my own, I would be a mother to all helpless children I could aid, and cure the pain at my own heart by soothing the pain of others."[33]

Erasmus Darwin, Charles Darwin, Joseph Dalton Hooker, and Thomas Henry Huxley had all urged work as a great balm to grief. It was an easier thing to prescribe to a man and a more common thing to expect, given Victorian gender roles. Besant rebelled, however, against what was demanded of her by middle-class Victorian society. Though heartsick, she "plunged into work with added vigour, for only in that did I find any solace." She admitted the pamphlets against Christianity that she wrote in this period were "marked by considerable bitterness," but Christianity, she explained, had robbed her of her child, and she struck mercilessly in return. To do so she took up a rigorous study of science, as she explained, to render herself more useful to the causes to which she had given her life. "The intellectual comprehension of the sources of evil," she wrote, "and the method of its extinction was the second great plank in my ethical platform." The study of Darwin, Huxley, Haeckel, and others not only convinced her of the "truth of evolution" but of "the evolution of the social instinct" as the only means of further progress. And for a time, at least, she believed atheism the only route toward full acceptance of the ethical stance that followed: "Not dreaming of a personal reward hereafter, not craving a personal payment from heavenly treasury, [the atheist] works and loves, content that he is building a future fairer than his present, joyous that he is creating a new earth for a happier race." In other words, the belief that natural laws, *and nothing else*, governed the fall of sparrows was the only belief that would save the world.[34]

Annie Besant did not remain an atheist. After 1889 she became a theosophist, one of various "spiritualist" movements strongly influenced by Indian philosophies. She also renounced her support for contraception after deciding that true human health and progress required a denial of animal instincts, including sex outside of procreation. After moving to India in 1898, Besant became a firm proponent of Indian self-rule and served as president of the Indian National Congress in 1917. (Besant had criticized Britain for its "brutal and brutalizing imperialism" since the 1870s.) There, she worked with Mohandas Gandhi on campaigns to ensure a better future for India's children. Although they disagreed on the best strategy for securing Home Rule, Besant and Gandhi both believed that the fact something existed beyond the material world need not halt action to change this one. Both campaigned at great personal risk to end British imperialism (Besant was also imprisoned in India, as, of course, was Gandhi). And both fought attempts to naturalize human conflict between either individuals (in, for example, capitalist economics) or groups (in imperialist ideologies) on the grounds these kinds of conflicts did *not* lead to progress or the amelioration of suffering for all children.[35]

What Strange and Subtle Influences

As Victorians debated whose children mattered most, they also debated how to save children and on what grounds. The stakes in these debates were clearly high, and they took place, of course, without the benefit of hindsight. Although child mortality rates (the number of children out of a thousand who died under the age of five) were still high in 1875, they had come down since the start of the Victorian era from about 300 to about 250 (today the number is four in Britain). The decline was not consistent (between 1895 and 1900 the number actually went up again), and the precipitous drops during which tuberculosis, scarlet fever, measles, diphtheria, and other infectious diseases disappeared from the top ten causes of child mortality did not begin until after 1900. In the meantime, the best means of lowering child mortality rates was not necessarily clear. Indeed, demographers still debate the degree to which sanitation and public health measures, as opposed to increases in living standards, moved mortality rates prior to 1900. With the exception of the smallpox vaccine, which had been in use for generations, vaccination played no role until the second quarter of the twentieth century, finishing off diseases like measles, diphtheria, and pertussis that had been declining for decades. (No one knows why scarlet fever, for which no vaccine exists, ultimately declined.)

Although the numbers were, in retrospect, going slowly and inconsistently down, children still ran a terrible gauntlet of infectious disease, as the constant threat of measles, smallpox, diphtheria, whooping cough, tuberculosis, cholera, typhoid, and other illnesses shows. On April 24, 1874, Alfred Russel Wallace and his wife Annie experienced this fact firsthand, for on that day, their firstborn son, Bertie, died shortly before his seventh birthday, apparently from complications of scarlet fever.

Unlike Annie Besant, Wallace's anguish in the face of this loss did not lead to a repudiation of a spiritual realm. He was grateful when his friend Arabella Buckley, a fellow Spiritualist, wrote to him of the comfort Spiritualism offered in such moments. "Poor Mrs Wallace," she wrote. "I am so very very sorry for her to be separated from her little man. I am afraid she will fret for him, & yet she is happy in knowing she has the power of hearing him, if not from him, whenever she cultivates it.

How wonderful it is how *completely* Spiritualism alters one's idea of death." Buckley reminded Wallace of the many friends who could get information for him on Bertie's doings in the spiritual realm. "I suppose Mrs Guppy having known dear little Bertie would be able to learn a good deal." (Mrs. Guppy was a well-known Spiritualist medium.) She wondered who would now take care of Bertie and "educate him for you." The question was important, for Spiritualists believed that, given that one's behavior on Earth determined the state of one's existence in the spirit world, those taken prematurely had none of the benefits of the "knowledge, attainments and experience of earth-life." Indeed, as Wallace explained in his book *A Defense of Spiritualism*, the fact one could continue to improve in the afterlife was one of Spiritualism's most important advantages over the system of punishments and rewards in orthodox Christianity: for a Spiritualist, everyone's misdeeds in this life could be slowly redeemed via a spiritual education in the afterlife. (Buckley soon reported that the spirits wished to inform Wallace that his brother, Herbert, who had died in the Amazon years earlier, would be taking Bertie's education in hand.)[36]

Wallace's vision held out hope for not only individual progress but the progress of humanity as a whole, both on Earth and in the hereafter. It also, he believed, served as a compelling solution to the problem of evil, since a benevolent purpose to suffering could be found in the fact that struggle led to progress. As we have seen, Darwin agreed that the struggle for existence had led to progress, but he refused to put a spiritual, benevolent force behind that fact. Wallace, by contrast, believed in a spiritual power in the universe, the presence of purpose and meaning within affliction, and material and moral progress as the ultimate goal toward which all affliction, *by design*, pushed human beings. He was, in other words, intent on recovering belief in a purposeful power within the laws governing the fall of sparrows. And in doing so, he recovered purpose and meaning in otherwise seemingly meaningless evolutionary answers to the question, "Why do some die and some live?" (This was the question that had first inspired his own discovery of natural selection.) When Bertie died, he targeted his despair toward doubts regarding the medical treatment Bertie had received, rather than the Ruler of the universe. Writing to his friend Arabella Buckley, Wallace lamented: "Our orthodox medical men are profoundly ignorant of the subtle influences of the human body in health and disease, and can thus do nothing in many cases which Nature would cure if assisted by proper conditions." He then added: "We who know what strange and subtle influences are around us can believe this."[37]

Darwin was hard at work revising *The Descent* and responding to his critics when Bertie Wallace died. When Wallace received the second edition of Darwin's book eight months later, he remained unmoved by Darwin's arguments that the human mind could have developed through purely naturalistic means. Wallace insisted that it was evidence, not grief or a wish for an afterlife, that had convinced him and others of a spiritual realm, meaning and purpose to human existence, and life after death. Mr. Andrew Leighton of Liverpool, for example, "has seen a pencil rise of itself on a table and write the words: '*And is this world of strife to end in dust at last?*'" George Sexton, previously a skeptical investigator of spiritual phenomena, had received information from communications that no one but the dead could possibly have known: Thus "our dear departed ones made themselves palpable both to feeling and to sight."[38]

Although Thomas Henry Huxley withheld his famous "beak and claws" from Wallace, he once wrote that he could not imagine a fate worse than the existence Spiritualists imagined for themselves and their loved ones after death. "The only good that I can see in the demonstration of the truth of 'Spiritualism,'" he quipped, "is to furnish an additional argument against suicide. Better live a crossing-sweeper than die and be made to talk twaddle by a 'medium' hired at a guinea a *séance*." Wallace stuck to his guns, despite the fact he knew Spiritualism placed him in the category of "crank and a faddist" by friends he respected and loved. He answered the "caustic satire" of a "kind-hearted Professor" (Huxley) by noting that those who had died at a more advanced state of intellectual and moral progress probably did have better things to do than talk to mediums.[39]

In the spring of 1878, Thomas and Nettie Huxley also had to confront the threat of pervasive child disease, yet again, when one of the most dreaded childhood illnesses, diphtheria, entered the Huxleys' home.

Huxley recounted how "two of the cases were light, but my Madge suffered terribly, and for some ten days we were in sickening anxiety about her." Nineteen-year-old Madge temporarily lost use of her sight, speech, and legs, her throat was in agony, and she was gasping and feverish, while "the family doctor stood by helpless." His friend T. J. Parker described the impact of his daughter's illness on the famously stoic Huxley: "I never saw a man more crushed than he was during the dangerous illness of one of his daughters, and he told me that, having then to make an after-dinner speech, he broke down for the first time in his life, and for one painful moment forgot where he was and what he had to say."[40]

Huxley recovered himself, and set to work applying new discoveries about "how nature works" to halting the disease. For, just eighteen years after his firstborn son's death, he believed science had produced clear *action* to be undertaken in the wake of an epidemic. By 1870, Louis Pasteur and Robert Koch had secured a hearing for the germ theory of disease. (Theories of contagion and infection had been around for centuries, but precisely *what* moved around was hotly debated.) Huxley was one of the first to adopt the new germ theory, sending vials of "bacilli" back and forth to his friend, the physicist John Tyndall. As word of Pasteur and Koch's work spread, physicians wondered: Was this why Joseph Lister's antiseptic procedures, first used in 1865, prevented postsurgical infection? Was the carbolic acid killing minute germs that caused all these terrible contagious diseases? Were the culprits of John Snow's proposal, in 1854, that cholera spread via contaminated water in fact microscopic germs? Meanwhile, as scientists debated germ theory throughout the 1870s, the fact it mapped well onto existing ideas about sanitation heightened calls for the prevention of disease by ensuring children received pure milk and clean water.[41]

To Huxley, germ theory represented a purely naturalistic explanation, rooted in experimental sciences, for why children became sick and died. Certain that the methods and assumptions that led to such discoveries were the only means of figuring out how to lower child mortality in future, Huxley and his students laid the groundwork for broad state support for his "New School of Prophets." They campaigned for a system of medical education rooted in laboratory-based biological research, and staffed Royal Commissions to study infectious diseases in order to figure out how to halt epidemics. They promised a great deal with, at first, little evidence that their methods could heal (as opposed to prevent) children's ailments. (Pasteur developed the first rabies vaccine in 1885, but a large-scale ability to cure disease via experimental medicine did not occur, with one important exception examined in the next section, until the next century.)

Having lost beloved children, both Huxley and Wallace believed that children could be saved in future via a better knowledge of nature. But they disagreed profoundly on what "a better knowledge of nature" meant and required. Huxley was a firm proponent of compulsory smallpox vaccination. Wallace, by contrast, argued against compulsory vaccination on the grounds it allowed the governing classes to ignore the suffering of the poor. He was certain that malnutrition influenced susceptibility to smallpox (healthy children, he suspected, did not succumb to the disease), and he hated the fact that compulsory and punitive vaccination laws primarily affected the poor. Wallace ridiculed the idea that vaccination could take the place of establishing healthier environments for children. He knew that once those in power imagined that disease could be prevented by vaccination, they would decide that nothing needed to be done to improve the environments in which the poor lived. Clearly, some of the foremost evolutionists, while they agreed that the study of nature's laws produced a more accurate account of why children died, did not agree on the best means of changing the world. These debates were heightened, rather than dampened, when science finally delivered on its promise to, as Huxley wrote to Kingsley, "work miracles."[42]

The Only One That Can Work Miracles

Soon after Madge's illness, Huxley organized an enquiry to establish the source of the diphtheria outbreak. He began the enquiry's first public meeting by insisting that although "the misery and suffering which had occurred was not traceable to any visible agency . . . there was no doubt that the course of the disease was perfectly definite and traceable, and quite capable of prevention if proper sanitary precautions were taken." Huxley's enquiry eventually attributed the London outbreak to the "culpable negligence" of those who had installed a small drain pipe. And the appropriate means of preventing future suffering was to simply fix the pipe.[43]

It turns out that Huxley was wrong about the pipe as the source of the outbreak, but developments were in motion that, within a generation, would hand physicians the means of not just preventing diphtheria's spread but, for the first time in history, actually snatching children from its clutches once the bacteria had made a child ill. In 1883 the German-Swiss pathologist Edwin Klebs and German bacteriologist Friedrich Loeffler identified the diphtheria bacillus and proved it caused the disease. (Every demonstration of a bacterial cause for a disease required producing that disease in experimental animals.) Meanwhile, the French bacteriologist Emile Roux showed that one could remove the bacteria, yet the terrible symptoms remained, thereby demonstrating that a toxin produced by the bacteria was what actually strangled children.

The discovery that exposure to diphtheria toxins conferred subsequent immunity led to hopes that scientists might harness the body's response to produce "antitoxins." Following experiments on animals, the first use of diphtheria antitoxin in humans occurred in 1891 in Berlin. By 1896 the Glasgow Fever Hospital reported that administration of antitoxin was the most likely cause of a reduction in mortality rates among diphtheria patients from 35.5 percent to 14 percent. With improved information about dosage and timely delivery, the rates dropped still further. By the end of the decade, antitoxin represented the first robust vindication of long-standing hopes that one day science would allow physicians to save lives *after* so terrible a disease had lodged in the body. Physicians could now conquer diphtheria itself, rather than merely try to prevent the bacillus from spreading or waiting helplessly for nature's "healing powers" to do their work (and sometimes fail) after a child fell ill. And it had all come about through the careful study of uniform cause and effect in a laboratory setting.[44]

Historian of science Evelynn Hammonds notes that, thanks to antitoxin, diphtheria was "the first infectious disease to be controlled by advances in scientific medicine, particularly discoveries in bacteriology and immunology."[45] We have seen how, in his letter to Reverend Charles

Kingsley after the death of Noel Huxley in 1861, Huxley had warned Kingsley that if "that great and powerful instrument for good and evil, the Church of England" was to survive the advancing tide of science, Anglican leaders must acknowledge the universality of nature's laws. Science, Huxley argued, was the only "school of the prophets . . . that can work miracles, the only one that can constantly appeal to nature for evidence that it is right." Huxley had used an old word, *miracles*, for what he hoped and believed science might do, including saving children so that they might spend a longer time on Earth. At the time he wrote to Kingsley, science had little to show for such hopes. But with diphtheria antitoxin, the biological sciences and laboratory methods had finally delivered their first miracle, over two centuries after Francis Bacon first promised that science would serve for "the relief of man's estate."

Huxley died in 1895, just as hints that antitoxin might save lives were spreading. His students, as men trained to ground medicine in the biological and laboratory sciences, became some of the most ardent proponents of both vaccination and antitoxin. Clearly, they agreed that child mortality could and should be lowered through a better knowledge of nature's laws. However, one of the tragic ironies of this history is the fact that, pursued within a society willing to place human beings on hierarchies of "fitness," the ability to lower child mortality rates inspired anxiety in some that saving all children would interfere in nature's laws too much, and in doing so, even lead to the downfall of "civilization." Darwin's version of this anxiety appeared, albeit briefly, in *The Descent of Man*. Writing on the question of whether natural selection still operated within "civilized" societies, Darwin wrote: "We build asylums for the imbecile, the maimed, and the sick; we institute poor-laws; and our medical men exert their utmost skill to save the life of every one to the last moment. There is reason to believe that vaccination has preserved thousands, who from a weak constitution would formerly have succumbed to small-pox. Thus the weak members of civilised societies propagate their kind." No one, he wrote, who had attended to the breeding of domestic animals "will doubt that this must be highly injurious to the race of man." Given his belief that "man's present rank" had arisen from a struggle for existence,

Darwin thus expressed some ambivalence about oft-touted hallmarks of medical progress.[46]

Darwin had immediately followed these lines with a warning that, given the importance of the instinct of sympathy to the evolution of the human moral sense, refusing "the aid which we feel impelled to give to the helpless," even "if so urged by hard reason" could not be done without "deterioration in the noblest part of our nature." His memories of pre-anesthetic surgery pervaded his explanation of why human beings must not, in fact, be treated like barnyard animals: "The surgeon may harden himself whilst performing an operation," he wrote, "for he knows that he is acting for the good of his patient; but if we were intentionally to neglect the weak and helpless, it could only be for a contingent benefit, with an overwhelming present evil." Clearly, Darwin was not opposed to speaking of "the undoubtedly bad effects of the weak surviving and propagating their kind," but he assumed that commitments to the higher values of both sympathy and individual liberty would prevent men from applying the "principles of the barnyard" to human beings.[47]

His assumption was wrong. In 1883, a year after Darwin's death, Darwin's cousin, Francis Galton, coined the word *eugenics* (Greek for "of good birth") to capture the possibility of guiding the fate of human populations now that the principles of natural selection had been discovered. In defending eugenics, Galton turned Darwin's warning about the danger of eugenic thinking to human sympathy on its head by describing eugenics as a kinder, more sympathetic replacement of nature's ways, writing: "Man is gifted with pity and other kindly feelings; he has also the power of preventing many kinds of suffering. I conceive it to fall well within his province to replace Natural Selection by other processes that are more merciful and not less effective. This is precisely the aim of Eugenics." Speaking to the Sociological Society in 1904, Galton urged that eugenics "must be introduced into the national conscience, like a new religion. . . . What nature does blindly, slowly, and ruthlessly, man may do providently, quickly, and kindly."[48]

Galton's "new religion" took root in part because, amid an increasing perception that infant and child mortality rates were dropping, some wondered whether the project of amelioration might *increase* human suffering (thus altering the assumed progressive direction of history) by saving the "wrong" children. A complex mix of hope of progress in the amelioration of suffering with anxiety regarding "race degeneration" (some meant the human race, others only their own) undergirded the broad support for eugenic thinking in Britain and elsewhere. Many of the most influential proponents of the new fields of pediatrics and public health feared that saving all babies and then letting them reproduce as adults might halt all progress in the amelioration of suffering. After all, since the seventeenth century, proponents of mechanistic science believed that the study of uniform natural laws would ameliorate suffering in future. Galton's vision of eugenics simply applied that tradition to new knowledge about heredity. Confident in the ability of scientists and physicians to declare what constituted both fitness and suffering, proponents of eugenics (including Anglican clergymen like William Inge, Dean of St. Paul's Cathedral from 1911 to 1934) saw themselves as fulfilling Francis Bacon's centuries-old promise to apply the study of nature's laws to the "relief of man's estate." At least for some.[49]

Versions of eugenic thinking existed across the political spectrum and almost the entire gamut of religious stances. Alfred Russel Wallace declared eugenics "simply the meddlesome interference of an arrogant scientific priestcraft" if practiced in a *capitalist* society. "He feared," recounted a friend, "that, as he understood it, Eugenics would perpetuate class distinctions, and postpone social reform, and afford quasi-scientific excuses for keeping people 'in the positions Nature intended them to occupy.'" But he was quite comfortable with the idea that women could and should choose their mates on eugenic grounds in a *socialist* society. "Change the environment," he argued, "so that all may have an adequate opportunity of living a useful and happy life, and give woman a free choice in marriage; and when that has been going on for some generations you may be in a better position to apply whatever has been discovered about heredity and human breeding, and you may then know which are the better stocks."[50]

Others thought capitalism did a fine job of sorting out the "better stocks," although they disagreed regarding the best policies for supporting that "sorting." Some, for example, argued against programs aimed at ameliorating the high child-mortality rates of the "inferior" classes. The British statistician (and Galton's protégé) Karl Pearson, who directed University College London's Galton Laboratory for National Eugenics, suggested that efforts to lower infant mortality would interfere in the "beneficent action" of natural selection. By contrast, the Edinburgh physician Caleb Saleeby (who coined the term *eugenist*) repudiated the suggestion that high infant mortality must be tolerated as the benevolent means of weeding out the "unfit." On moral grounds, Saleeby insisted, the eugenist must do his best "for all children, once they are born—nay, more, from the very moment, months before, at which their individual history starts." But he agreed that the fight to lower infant mortality would cause more suffering within the century, *if not accompanied by eugenic measures.* The unfit must be given sympathy, he concluded, but not the right of parenthood.[51]

Saleeby connected his policy recommendations explicitly to Darwin's explanation of the world. In 1909, the centenary of Darwin's birth, Saleeby described his own work as "a humble tribute to that immortal name, for it is based upon the idea of *selection for parenthood* as determining the nature, fate and worth of living races, which is Darwin's chief contribution to thought, and finds in eugenics its supreme application." Saleeby included everything from the prevention of alcoholism to sanitation reform as within the purview of eugenics. "Our age is now awakening, at least, to the cry of the children," he wrote, but in answering that cry, all who love children and who are acquainted with the principles of organic evolution must "teach and preach" the main lesson of eugenics, namely, "the rejection of the unworthy, *not as individuals but as parents.*" Without this "scientific" guidance of humanitarian action, Saleeby warned, "we shall assuredly breed for posterity, whose lives and happiness and moral welfare are in our hands, evils that can adequately neither be named nor numbered."[52]

This was clearly not the kind of scientific miracle-working Thomas Henry Huxley had in mind when he warned Reverend Kingsley that the Anglican Church best get out of the way of science. In an 1894 essay entitled "Evolution and Ethics," Huxley ridiculed the idea that some "scientific administrator" of society might be able to select "only the strong and the healthy" to have children. Even the keenest judge of character, Huxley countered, could not "pick out with the least chance of success" who should be allowed and who should be prevented from having children. Indeed, in contrast to those who argued against eugenics on the grounds scientists did not yet know enough about heredity (thus implying approval of eugenics so long as the science was better), Huxley was absolutely certain that "there is no hope that mere human beings will ever possess enough intelligence to select the fittest." He had too little confidence in men's judgments of their fellows and too much faith in the capacity of individuals to improve. "Surely, one must be very 'fit,' indeed," he quipped, "not to know of an occasion, or perhaps two, in one's life, when it would have been only too easy to qualify for a place among the 'unfit.'" Ironically, as we saw above, in the 1860s Huxley was quite willing to declare non-Europeans "by nature" less fit for modern civilization. Yet his distrust in the ability of human beings to decide some children were worth more than others, no matter how much science they knew, had the capacity to halt both eugenic and racist thinking in its tracks.[53]

To some extent, Darwin shared Huxley's skepticism regarding the ability of human beings to make the kinds of judgments eugenists presumed possible. When, in 1873, he read Galton's proposal that a "register" be kept of "naturally gifted individuals," he told Galton that he thought "the greatest difficulty" would be "in deciding who deserved to be on the register." In *The Descent of Man*, the most he was willing to suggest was that "both sexes ought to refrain from marriage if they are in any marked degree inferior in body or mind." Yet he also thought that, while "such hopes are Utopian," they would "never be even partially realised until the laws of inheritance are thoroughly known." Clearly, and unlike Huxley, Darwin thought improved knowledge about heredity might make certain kinds of mating decisions both possible and valuable in future. Indeed, he then added the following uncharacteristically passionate line: "When the principles of breeding and inheritance are better understood, we shall not hear ignorant members of our legislature rejecting with scorn a plan for ascertaining whether or not consanguineous marriages are injurious to man."[54]

Immersed in the study of variation, Darwin clearly wondered whether first-cousin marriages compromised the health of progeny, including his own. (Of course, given the fact almost half of British children under the age of fifteen died in this era, Emma and Charles's experience was hardly unique, cousin-marriage or no.) In 1870 he had lobbied for a question about such marriages to be added to the national census so that answering "whether or not consanguineous marriages are injurious to man" could be based on evidence rather than anecdotes. The House of Commons voted down the proposition on the grounds that the government should not be involved in determining who should be allowed to marry whom. But Darwin's willingness to even ask the question demonstrates his betimes heartbreaking discipline in following out a thought no matter where it might lead: if cousin-marriage led to unhealthy offspring, neither Providence nor chance but his own choices amid ignorance of nature's laws may have led to the death of three of his children.

By the time Darwin posed this question about the influence of first-cousin marriages on the health of children, the words of William Owen (with which we opened this book) that "all is for the best & that the blow has been struck in mercy by that Almighty Spirit who we are bound to believe cannot err, & without whose knowledge & will not a Sparrow falls" meant nothing to him.[55] Natural law, and natural law alone, governed each sparrow's fall. Charles Darwin was not alone, of course, in deciding that Owen's belief in a close personal God who governed the fall of every sparrow did not make adequate rational or ethical sense of the world. On the other hand, not all who adopted Darwin's evolutionary explanation of the world thought that doing so required abandoning belief in a close personal God. But whether they tried to resign themselves to God's will, nature's laws, or some complex mix of the two, clearly, for those sitting "beside the death-beds of amiable children," all options demanded a great deal from the human heart.[56]

Conclusion

At the end of July 1914, European powers punctuated two centuries of promises that a better knowledge of nature's laws would ameliorate suffering by descending into war. For four years, Britain and its imperial possessions, France, and Russia (and eventually the United States) fought Germany, Austria-Hungary, and Italy. Before the war ended in 1918, nine million men brought through the perils of diphtheria, measles, and scarlet fever as children were blown to bits, mowed down by machine guns, gassed in the trenches, or struck by infectious diseases. All sides, from Russian Bolshevists to British biologists, called upon evolution to both explain the war and imagine how society should be organized when the war ended. They would repeat those arguments, with even more terrible weapons, in 1938, after Nazi Germany turned on its own people under cover of "race hygiene" and began a merciless struggle to expand "living space" for "the Aryan race."[1]

After World War I, the American biologist Vernon Kellogg (who served with the Red Cross in Belgium) published an account of imprisoned German military officers' justifications of the war. Kellogg had spent hours listening to a military officer, also one of the foremost biologists in Germany, defending the war. Darwin, the officer claimed, had demonstrated that civilization progresses solely via a violent struggle between groups. Thus, this German biologist embraced Darwin's emphasis on the successful prospects of groups who had a high degree of sympathy and loyalty to those within their own group, but repudiated Darwin's conclusion that "the simplest reason" would inspire men to extend those sympathies outside their group. In his own writings, Kellogg argued, by contrast, that evolution showed that increased cooperation and mutual aid *between* groups led to progress. Kellogg thus insisted on Darwinian grounds that war was both scientifically and morally wrong. Both sides in this debate drew upon evolution to make arguments about what the world should be like, much like individuals have also long drawn upon the Bible to defend diametrically opposite positions.[2]

Having read Kellogg's book about German biologists' use of evolution to justify war, in 1921 three-time US democratic presidential candidate and former Secretary of State William Jennings Bryan diagnosed the world's recent troubles as rooted in Darwinian explanations of the sparrow's fall. Imagine, he wrote, that a child grows up in the folds of a Christian household: "He talks with God. He goes to Sunday school and learns that the Heavenly Father is even more kind than earthly parents; he hears the preacher tell how precious our lives are in the sight of God—how even a sparrow cannot fall to the ground without His notice." He learns, Bryan wrote, of God's benevolent governance over every individual life, and that every joy and every affliction reflects the will and purposes of a wise Creator who can be trusted. Then the child goes to college and "some learned professor" leads him through six hundred pages of Charles Darwin's *The Descent of Man*, where he learns the Bible stories are myths; he has a lineage millions of years old; and he is kindred to apes, rather than Adam. Bryan demanded to know how America's children, after being taught such things, could still feel God's presence in their daily lives. How could they still believe and trust that "even a sparrow cannot fall to the ground without His notice"?[3]

To Bryan, a correct understanding of human origins and history depended upon faith in a personal God who attended to the fall of every sparrow and for whom one miracle was as easy as any other. In 1904 Bryan spoke of his fear that "we shall lose the consciousness of God's presence in our daily life, if we must accept the theory that through all the ages no spiritual force has touched the life of man and shaped the destiny of nations." After the war, he believed he had plenty of evidence that abandoning these beliefs for the sake of some evolutionary visions of creation, whether under God's governance or not, caused tremendous suffering. For Bryan, the lesson for what must be done to end such violence and improve the earth was clear: recover a sense of God's close providence in the lives of individuals and the nation. Some, of course, had found that distancing God from creation made better sense of the world. Some even argued that doing so inspired individuals to imagine more scope for human action. But Bryan, while he agreed humans could and must work to improve the human condition, was certain that such efforts must be grounded in Christianity. From the war to eugenics, he took the present and recent past as the best evidence of what might otherwise go so terribly wrong.[4]

Convinced that Darwinism posed tremendous dangers to the faith of American youth and the social and moral order of society, Bryan began a campaign in 1921 to get evolution out of US public schools. In books, speeches, and newspapers, he delivered passionate warnings of the consequences of teaching Darwinian theory to children. His campaign culminated in the famous Scopes "Monkey" Trial of 1925, when high school teacher John T. Scopes volunteered to test a Tennessee law that made it a crime to teach "any theory that denies the Story of the Divine Creation of man as taught in the Bible, and to teach instead that man has descended from a lower order of animals." Scopes's defense team brought in Christian ministers, geologists, and biologists, and Reform Jewish rabbis to testify that evolution could be reconciled with belief in God. But Bryan's argument that evolutionary explanations of the world are the source of everything wrong with the world continues to be echoed by those opposed to teaching evolution in schools, especially in the United States and Turkey.[5]

Stories about Darwin have been used as ammunition in fights over teaching evolution ever since Bryan launched his campaign. As evidence that evolution destroyed faith in a close personal God, for example, Bryan included the lines from Darwin's own autobiography in which Darwin questioned the miracles of Christianity. Darwin's own descent into unbelief, Bryan insisted, clearly demonstrated the "*natural tendency* of Darwinism": the replacement of Christ's miracles and promise of heaven with natural laws that cared nothing for human beings. He quoted Darwin's words that science had nothing to do with Christ, except "in so far as a habit of scientific research makes a man cautious in admitting evidence." And he included Darwin's doubt regarding whether the mind of man can be trusted when imagining God if humanity came from apes. Based on Darwin's own account, Bryan insisted that the doctrine of evolution had led Darwin and millions of others astray from Christianity, with tragic results for individual souls and civilization itself. He then called upon American parents, legislators, and school boards to halt that "natural tendency," starting with the nation's schoolrooms.

Though he explored the complex landscape of Victorian beliefs, Bryan left out key elements of why Darwin's ideas about God and evolution changed over time. Darwin had actually included many reasons for his descent toward unbelief, and some of the most important had very little to do with either science or evolution. To be fair, thanks to Emma's editing, Bryan did not have access to Darwin's indignant lines about the "damnable" doctrine that unbelievers went to hell. But Bryan did have access to Darwin's lines about two beliefs that he chose not to mention: first, Darwin's suspicion that the existence of many other religions undermined Christianity's claim to be a unique revelation; and second, Darwin's passionate rebellion against Christian explanations of why so much suffering exists in the world. The first involved knowledge that had nothing to do with evolution, but arose from Europeans' exposure to Islam and (at least Victorian concepts of) Hinduism, Buddhism, and other non-Christian belief systems. The second concerned ethical challenges to belief in a close personal God that one can find in the ancient texts of many philosophical and religious traditions. The problems, in other words, were old; they did not suddenly appear with the publication of either *On the Origin of Species* or *The Descent of Man*. Of course, the fact that both these challenges existed long before the discovery of evolution did not map well onto Bryan's argument that Darwinism led children astray from Christian beliefs. Bryan (and present-day critics of evolution) needed evolution to be the cause of disbelief, rather than a potential consequence of doubt that arose from other quarters.

Of course, tracing the trajectory of Darwin's (or anyone else's) stance on evolution and/or God is a complicated proposition. In following the thread of child loss, this book has highlighted the role played by Victorians' explanations of suffering, but it is important to remember that a range of factors influenced what Darwin and his fellow Victorians

thought and believed about both nature and God. From fascinating natural history puzzles and changing concepts of science, to the debate over
slavery and exposure to non-Christian religious traditions, many different
yet interweaving threads convinced Darwin that evolution explained
both the origin of species and human experience better than the theory
of special creation. Indeed, Darwin knew that convincing other naturalists depended on whether his theory explained a whole range of natural
phenomena, and not just parasitic wasps. Additionally, watching Victorians wrestle with ideas about God and nature amid high child mortality
rates illustrates how decisions about evolution have always entailed much
more than a simple comparison of Darwin's theory with the fossil record,
rudimentary organs like the human appendix, or the beaks of Galapagos
finches. After all, Darwin reserved an explicit statement that evolution
also explained the pervasive amount of suffering in the world for a doc-
ument composed solely for his family.

Both sides of today's "evolution vs. creation" debates have tried to use
Darwin's story to support their arguments. In 2001, biochemist and proponent of intelligent design (a twenty-first-century version of the theory
of special creation) Cornelius Hunter claimed that first, Darwin's theory
was primarily inspired by his rebellion against Christian explanations
of the existence of evil and suffering, and second, that this fact undermines the scientific credibility of evolutionary biology.[6] (This is a strange
argument to make, for it makes the truth content of a theory dependent
upon its history, rather than its ability to explain a wide range of facts.)
Meanwhile, defenders of evolution ignore the role played by Darwin's
changing assumptions about God by telling simplistic stories of Darwin
"discovering" natural selection while watching finches on the Galapagos
Islands. Such tales profoundly obscure the complex factors that have
influenced explanations of the origin of species, including *Homo sapiens*.[7]

Like William Jennings Bryan, defenders of intelligent design cite the
fact that Darwin's work was used to justify racism, eugenics, and war as
evidence against the theory of evolution. In response, many defenders of
evolution have tried to absolve Darwin's work of political taint by claiming his ideas were distinct from "social Darwinism." Historian Evelleen
Richards notes that often "Darwin is presented as the young naturalist of

the 'Beagle,' subsequent pigeon breeder and barnacle dissector and, above all, detached and objective observer and theoretician—remote from the political concerns of his fellow Victorians who misappropriated his scientific concepts to rationalize *their* imperialism, laissez-faire economics and racism."[8] Indeed, historians James Moore and Adrian Desmond note that the persistence of the view that "Darwinism was science, Social Darwinism ideology, and never the twain should meet" is good evidence that "*The Descent of Man* remains Darwin's greatest unread book."[9] Careful historical analysis reveals that, in fact, few naturalists, including Darwin, were willing to give up nature as a source for social prescription. As often as not, fights over the best explanation of the origin, purpose, and response to suffering, including child mortality, were the threads that bound explanations of nature and society together.

Today, the problem of suffering rarely explicitly appears in biology textbooks. But, given how the history of evolutionary thinking has been intertwined with attempts to understand the origin, purpose, and best response to suffering, it is not surprising that such attempts continue to exist between the lines of stances and debates. Sometimes they even come to the surface. The famous naturalist and broadcaster David Attenborough (an agnostic) was once asked whether he ever railed against nature's cruelty. "That assumes it has a choice," Attenborough replied, "in the sense that kindness is a positive act. Nature isn't positive in that way. It doesn't aim itself at you. It's not being unkind to you. A tsunami is the force of gravity, something to do with matter, what the world is." (The example was personal: Attenborough lost his niece and grandniece to the Indian Ocean tsunami in 2004.) When asked why he did not mention God in his nature shows, Attenborough replied that people tend to mention hummingbirds as examples of the wonders of God's creation, but "I tend to think of an innocent little child sitting on the bank of a river in Africa, who's got a worm boring through [his] eye that can render him blind before he's eight. Now, presumably you think this Lord created this worm, just as he created the hummingbird. I find that rather tricky."[10]

Ultimately, this history reminds us that assessing claims about the origin of species has always entailed much more than deciding the best explanation of finch beaks. This history demonstrates how difficult it has always been to disentangle claims about either nature or God from the search for purpose and meaning that accompany moments of bliss or misery. Although religion is often described as deriving power from the purpose it confers on suffering, the story of Victorian evolutionists shows that science, too, derived power from the same source, long before it could do anything to ameliorate that suffering. Finally, this history shows that it is impossible to assess the legacies of Darwin's emphasis on natural law as either solely good or solely evil. We have inherited *both* tremendous tragedies (for example, extraordinary inequities of child disease and mortality around the world) and astonishing triumphs (for example, the ability to halt infectious diseases like river blindness in their tracks with antibiotics and antiparasitics) from this complicated past.

NOTES

INTRODUCTION

1. William Owen to Charles Darwin, March 26, 1843, Darwin Correspondence Project (DCP), "Letter no. 666."

2. Philip Doddridge, *Submission to Divine Providence in the Death of Children, recommended and inforced, in a Sermon preached at Northampton, On the Death Of a very amiable and hopeful Child, about Five Years old. Published out of Compassion to mourning Parents* (London: R. Hett, 1737), 32.

3. For scholarship that undermines Philip Aries's thesis, see Hannah Newton, *The Sick Child in Early Modern England, 1580–1720* (Oxford, UK: Oxford University Press, 2012), and Nicholas Orme, *Medieval Children* (New Haven, CT: Yale University Press, 2001).

4. E. Harold Browne, *An Exposition of the Thirty-Nine Articles*, 6th ed. (London: Longman, Green, Longman, Roberts & Green, 1864), 13.

5. See David N. Livingstone, *Darwin's Forgotten Defenders: The Encounter between Evangelical Theology and Evolutionary Thought* (Vancouver: Regent College Publishing, 1997), and James R. Moore, *The Post-Darwinian Controversies: A Study of the Protestant Struggle to Come to Terms with Darwin in Great Britain and America, 1870–1900* (Cambridge, UK: Cambridge University Press, 1981).

6. David C. Lindberg and Ronald L. Numbers, "Beyond War and Peace: A Reappraisal of the Encounter between Christianity and Science," *Church History* 55, no. 3 (1986): 340. For historical work that corrects some of the legends and myths commonly told about the history of science and religion, see Ronald L. Numbers, ed., *Galileo Goes to Jail and Other Myths about Science and Religion* (Cambridge, MA: Harvard University Press, 2010). For the problem of avoiding present-day definitions in studying the history of science and religion in the past, see John Hedley Brooke, *Science and Religion: Historical Perspectives* (Cambridge, UK: Cambridge University Press, 1991), 5–11.

CHAPTER ONE

1. John Ray, *The Correspondence of John Ray: Consisting of Selections from the Philosophical Letters Published by Dr. Derham, and Original Letters of John Ray, in the Collection of the British Museum* (London: Ray Society, 1848), 311–13.

2. See the discussion in James Turner, *Without God, without Creed: The Origins of Unbelief in America* (Baltimore: Johns Hopkins University Press, 1985), 26.

3. Hannah Newton, "The Dying Child in Seventeenth-Century England," *Pediatrics* 136, no. 2 (August 2015): 220.

4. Alice Thornton, *The Autobiography of Mrs. Alice Thornton, of East Newton Co. York* (Durham, UK: Andrews and Co., 1875), 126.

5. On Ray and natural theology more generally, see Neil C. Gillespie, "Natural History, Natural Theology, and Social Order: John Ray and the 'Newtonian Ideology,'" *Journal of the History of Biology* 20 (1987): 1–49.

6. On the influence of the English Civil War on the priorities of the Royal Society, see the classic account by Richard S. Westfall, *Science and Religion in Seventeenth-Century England* (New Haven, CT: Yale University Press, 1958).

7. See Steven Shapin and Simon Schaffer, *Leviathan and the Air-Pump: Hobbes, Boyle and the Experimental Life* (Princeton, NJ: Princeton University Press, 1989).

8. Brooke, *Science and Religion*, 118.

9. On Spinoza, see Steven M. Nadler, *Spinoza: A Life* (Cambridge, UK: Cambridge University Press, 2001).

10. John Ray, *The Wisdom of God Manifested in the Works of the Creation* (London: Samuel Smith, 1691), 18, 102, 105. For the classic biography of John Ray, see Charles Raven, *John Ray, Naturalist: His Life and Works* (Cambridge, UK: Cambridge University Press, 1942).

11. Ray, *The Wisdom of God*, 155, 156, 162. For Ray's influence on biology, see Ernst Mayr, *The Growth of Biological Thought: Diversity, Evolution, and Inheritance* (Cambridge, MA: Harvard University Press, 1982), 104.

12. St. Basil the Great, "Question 55" in "The Long Rules," *St. Basil: Ascetical Works*, trans. M. Monica Wagner, vol. 9 of *The Fathers of the Church: A New Translation* (Washington DC: Catholic University of America Press, 1962), 330–37. For the medieval Christian view, see Roy Porter, *The Greatest Benefit to Mankind: Medical History of Humanity from Antiquity to the Present* (London: Harper Collins, 1997), 129.

13. See Peter Harrison, *The Fall of Man and the Foundations of Science* (Cambridge, UK: Cambridge University Press, 2007). David Noble traced Christian visions of technology as a means of recovering Eden back to the medieval period in *The Religion of Technology: The Divinity of Man and the Spirit of Invention* (New York: Alfred A. Knopf, 1997).

14. On Francis Bacon and his influence on the Royal Society and new sciences, see Zachary McLeod Hutchins, "Building Bensalem at Massachusetts Bay: Francis Bacon and the Wisdom of Eden in Early Modern New England," *The New England Quarterly* 83 (2010): 601; Joanna Picciotto, "Reforming the Garden: The Experimentalist Eden and 'Paradise Lost,'" *English Literary History* 72, no. 1 (April 1, 2005): 23–78; and Michael Hunter, *Science and the Shape of Orthodoxy: Intellectual Change in Late Seventeenth-Century Britain* (Woodbridge, UK: Boydell Press, 1995), 14–16.

15. Ray, *The Wisdom of God*, 114–15 and, from the 1704 edition, vol. 1, 192, and vol. 2, 245.

16. Ray, *Correspondence*, 312.

17. Margaret Cavendish, *Observations upon Experimental Philosophy*, edited by Eugene Marshall (Indianapolis: Hackett Publishing, 2016), 8. From the 2nd edition (London: A. Maxwell, 1668).

18. See, for example, David Noble, *A World without Women: The Christian Clerical Culture of Western Science* (New York: Alfred A. Knopf, 1992); and Londa Schiebinger, *The Mind Has No Sex?: Women in the Origins of Modern Science* (Cambridge, MA: Harvard University Press, 1991). On Margaret Cavendish, see Anna Battigelli, *Margaret Cavendish and the Exiles of the Mind* (Lexington: University Press of Kentucky, 1998); and Lisa T. Sarasohn, *The Natural Philosophy of Margaret Cavendish: Reason and Fancy during the Scientific Revolution* (Baltimore, MD: Johns Hopkins University Press, 2010).

19. Phillis Wheatley, *Memoir and Poems of Phillis Wheatley, A Native African and a Slave* (Boston: Geo W. Light, 1834), 86.

20. Wheatley, *Memoir and Poems*, 85.

21. Ibid., 42.

22. Ibid., 75.

23. See Jeffrey Bilbro, "Who Are Lost and How They're Found: Redemption and Theodicy in Wheatley, Newton, and Cowper," *Early American Literature* 57 (2012): 561–89.

24. Francis Darwin, ed., *The Life and Letters of Charles Darwin*, vol. 1 (London: John Murray, 1887), 47.

25. William Paley, *Natural Theology: or, Evidences of the Existence and Attributes of the Deity, Collected from the Appearances of Nature* (London: R. Faulder, 1802), 1–4.

26. Paley, *Natural Theology*, 12, 19–20.

27. Paley, *Natural Theology*, 71. See also Neal C. Gillespie, "Divine Design and the Industrial Revolution: William Paley's Abortive Reform of Natural Theology," *Isis* 81, no. 2 (1990): 229.

28. See David Lindberg, "Galileo, the Church, and the Cosmos," in *When Science and Christianity Meet*, ed. David Lindberg and Ronald L. Numbers, 33–66 (Chicago: University of Chicago Press, 2003).

29. Paley, *Natural Theology*, 299.

30. Ibid., 68.

31. Ibid., 572, 191–92.

32. Ray, *The Wisdom of God*, 162; Paley, *Natural Theology*, 466.

33. Paley, *Natural Theology*, 33, 578–79.

34. Edmund Paley, *The Works of William Paley, D.D. with Additional Sermons, etc. etc. and a Corrected Account of the Life and Writings of the Author*, vol. 7 (London: C. and J. Rivington etc., 1825), 36.

35. The prayers may be found in William Paley, *The Works of William Paley, D.D. Archdeacon of Carlisle. With a Life of the Author*, vol. 3 (London: Thomas Tegg, 1825), 275–410.

36. Paley, *Natural Theology*, 574.

37. William Paley, *Moral and Political Philosophy* (New York: S. King, 1824), 60; Martin Lowther Clarke, *Paley: Evidences for the Man* (Toronto: University of Toronto Press, 1975), 36, 45.

38. Paley, *Natural Theology*, 574; G. W. Meadley, "Memoirs of William Paley, D.D.," in *The Works of William Paley*, vol. 1 (Boston: J. Graham, 1810), 75.

39. William Paley, *The Works of William Paley, D.D. Archdeacon of Carlisle* (Edinburgh: Peter Brown and Thomas Nelson, 1829), 256, 276–77.

40. Paley, *Natural Theology*, 527–28.

41. Paley, *Natural Theology*, 529. For Paley's concession, see Dan Lloyd Le Mahieu, *The Mind of William Paley: A Philosopher and His Age* (Lincoln: University of Nebraska Press, 1976), 85.

42. Paley, *Natural Theology*, 5, 532.

43. Ibid., 490.

44. Charles Darwin to John Lubbock, November 22, 1859, DCP, "Letter no. 2532."

45. See Gillespie, "Divine Design and the Industrial Revolution"; Sujit Sivasundaram, *Nature and the Godly Empire: Science and Evangelical Mission in the Pacific, 1795–1850* (Cambridge, UK: Cambridge University Press, 2005); and Janet Browne, *Charles Darwin: Voyaging* (Princeton, NJ: Princeton University Press, 1996), 129.

46. Paley, *The Works of William Paley* (1829), 49.

CHAPTER TWO

1. Erasmus Darwin to James Watt, January 6, 1781, in *The Collected Letters of Erasmus Darwin*, ed. Desmond King-Hele (Cambridge, UK: Cambridge University Press, 2007), 181–82.

2. Brooke, *Science and Religion*, 118–19.

3. Anonymous, "Letter of Dr. Darwin," *Gentleman's Magazine and Historical Chronicle* 104 (1808): 869.

4. See Norton Garfinkle, "Science and Religion in England, 1790–1800: The Critical Response to the Work of Erasmus Darwin," *Journal of the History of Ideas* 16, no. 3 (1955): 376–88; and Erasmus Darwin to Josiah Wedgwood, April 4, 1786, in *The Collected Letters of Erasmus Darwin*, 250.

5. Anonymous, "Letter of Dr. Darwin," 869.

6. Erasmus's poem appears in Elizabeth Scott, *Specimens of British Poetry: Chiefly Selected from Authors of High Celebrity, and Interspersed with Original Writings* (Edinburgh: James Ballantyne and Co., 1823).

7. Turner, *Without God, without Creed*, 44.

8. Kathleen Anne Wellman, *La Mettrie: Medicine, Philosophy, and Enlightenment* (Durham, NC, and London: Duke University Press, 1992), 275.

9. See Ashley Marshall, "Erasmus Darwin *contra* David Hume," *British Journal for Eighteenth-Century Studies* 30, no. 1 (2007): 89–111.

10. John Ray, *Three Physico-Theological Discourses*, 3rd ed. (London: William Innys, 1713), 453.

11. Paley, *Natural Theology*, 566.

12. Joseph Priestley, *The Theological and Miscellaneous Works of Joseph Priestley*, vol. 18 (G. Smallfield, 1787), 527.

13. Philip Doddridge, *Submission to Divine Providence in the Death of Children, recommended and inforced, in a Sermon preached at Northampton, On the Death Of a very amiable and hopeful Child, about Five Years old. Published out of Compassion to mourning Parents* (London: R. Hett, 1737), 8.

14. William B. O. Peabody, *The Loss of Children: A Sermon, delivered January 22, 1832* (S. Bowler, 1832), 10.

15. Philip Doddridge, *Sermons to Young Persons* (London: J. Fowler, 1735), 201–2.

16. Doddridge, *Submission to Divine Providence*, 9.

17. Adrian Desmond and James Moore, *Darwin: The Life of a Tormented Evolutionist* (New York: Norton and Company, 1991), 5.

18. Thomas Balguy, *Divine Benevolence Asserted; And Vindicated from the Objections of Ancient and Modern Sceptics* (London: 1803; originally published 1781), 20, 39; Thomas Lowndes, *Tracts, Political and Miscellaneous, in Prose and Verse* (London: 1827), 323.

19. Roy Porter, *English Society in the Eighteenth Century* (New York: Penguin, 1982), 207.

20. Erasmus Darwin to Matthew Boulton, April 5, 1778, and Erasmus Darwin to Albert Reimarus, May 20, 1756, in *The Collected Letters of Erasmus Darwin*, 150–51 and 27–29.

21. A. N. Williams, "Physician, Poet, Polymath and Prophet: Erasmus Darwin a New Window into Eighteenth-Century Child Health Care," *British Society for the History of Paediatrics and Child Health* 96, suppl. 1 (2011): A43.

22. King-Hele, ed., *The Collected Letters of Erasmus Darwin*, 441, note 4; Erasmus Darwin to James Watt, June 21, 1796, in *The Collected Letters of Erasmus Darwin*, 500.

23. Erasmus Darwin, *The Botanic Garden*, vol. 1 (London: J. Johnson, 1791), 90–91.

24. Erasmus Darwin to Robert Darwin, early October 1792, in *The Collected Letters of Erasmus Darwin*, 408.

25. Porter, *The Greatest Benefit to Mankind*, 260.

26. Erasmus Darwin to Robert Darwin, May 30, 1792, in *The Collected Letters of Erasmus Darwin*, 403.

27. Darwin to Watt, January 1, 1794, and June 6, 1794, in *The Collected Letters of Erasmus Darwin*, 427 and 439.

28. Both are quoted in Garfinkle, "Science and Religion in England," 379–80.

29. King-Hele, ed., *The Collected Letters of Erasmus Darwin*, 104, 152–53.

30. Erasmus Darwin to Robert Darwin, March 20, 1800; to Thomas Beddoes, January 17, 1793; to Robert Darwin, April 13, 1792, and September 4, 1791, in *The Collected Letters of Erasmus Darwin*, 389, 415, 401, 544.

31. Patricia Fara, *Erasmus Darwin: Sex, Science, and Serendipity* (Oxford: Oxford University Press, 2012), 23.

32. Darwin to Wedgwood, April 13, 1789, in *The Collected Letters of Erasmus Darwin*, 338.

33. Darwin, *The Botanic Garden*, 96.

34. See Jean Max Charles, "The Slave Revolt That Changed the World and the Conspiracy against It: The Haitian Revolution and the Birth of Scientific Racism," *Journal of Black Studies* 51, no. 4 (2020): 275–94.

35. Larry R. Morrison, "The Religious Defense of American Slavery before 1830," *The Journal of Religious Thought* 37, no. 2 (1980): 28–29.

36. On the different uses of Mitchell's work, see James Delbourgo, "The Newtonian Slave Body: Racial Enlightenment in the Atlantic World," *Atlantic Studies* 9, no. 2 (2012): 182–207.

37. Browne, *Charles Darwin: Voyaging*, 38.

38. Erasmus Darwin, *Zoonomia, or the Laws of Organic Life*, vol. 1, 2nd ed. (Boston: Thomas and Andrews, 1803), 397.

39. Darwin, *Zoonomia*, vol. 1, 507–8.

40. Ibid., 400–401.

41. Ibid., 400–402.

42. Garfinkle, "Science and Religion in England," 379.

43. Erasmus Darwin, *The Temple of Nature; Or, the Origin of Society: A Poem, with Philosophical Notes* (London: J. Johnson, 1803), 26.

44. Darwin, *The Temple of Nature*, 134.

45. On Boyle, see Westfall, *Science and Religion in Seventeenth-Century England*, 30–31; Darwin, *The Temple of Nature*, 131.

46. Darwin, *The Temple of Nature*, 131.

47. Charles Darwin to Asa Gray, May 22, 1860, DCP, "Letter no. 2814."

48. Darwin, *The Temple of Nature*, 141.

49. Ibid., 162.

50. Ibid., 12.

51. Ibid., 61.

52. Ibid., 120.

53. Garfinkle, "Science and Religion in England," 386.

54. Ibid., 382.

55. Ibid., 384.

56. On Cuvier, see Toby A. Appel, *The Cuvier-Geoffroy Debate: French Biology in the Decades before Darwin* (Oxford: Oxford University Press, 1987).

57. Mary Shelley, *Frankenstein, or the Modern Prometheus* (London: George Routledge and Sons, 1891), ix–x.

58. Julian Marshall, *The Life and Letters of Mary Wollstonecraft Shelley*, vol. 1 (London: Richard Bentley & Son, 1889), 242, 225–27.

59. Percy Bysshe Shelley, *A Refutation of Deism: In a Dialogue* (London: Progressive Publishing Company, 1890; originally published 1814), 24.

60. Marshall, *The Life and Letters of Mary Wollstonecraft Shelley*, 110.

61. On the various editions of Paley's work, see Gillespie, "Divine Design and the Industrial Revolution," 226. On the continued dominance of Paley's ideas, especially in establishment medical schools, see Adrian Desmond, *The Politics of Evolution: Morphology, Medicine and Reform in Radical London* (Chicago: University of Chicago Press, 1989).

Chapter Three

1. Darwin, *The Life and Letters of Charles Darwin*, 33.

2. Monro is quoted in Browne, *Charles Darwin: Voyaging*, 52.

3. Desmond, *The Politics of Evolution*, 5.

4. Ibid., 53.

5. Quoted in Richard W. Burkhardt, *The Spirit of System: Lamarck and Evolutionary Biology* (Cambridge, MA: Harvard University Press, 1995), 171–72.

6. Darwin, *The Life and Letters of Charles Darwin*, 38.

7. Browne, *Charles Darwin: Voyaging*, 83.

8. Georges Cuvier, "Cuvier's Biographical Memoir of M. de Lamarck," *The Edinburgh New Philosophical Journal* 20, no. 39 (1836): 15.

9. On Darwin's regret regarding his lack of experience dissecting, see Darwin, *The Life and Letters of Charles Darwin*, 36; for Coldstream, see John Hutton Balfour, *Biography of the Late John Coldstream* (London: John Nisbet and Co., 1865), 39–41.

10. Balfour, *Biography of the Late John Coldstream*, 7, 109, 178.

11. William Kirby, *On the Power, Wisdom and Goodness of God as Manifested in the Creation of Animals and Their History, Habits, and Instincts*, vol. 1 (London: William Pickering, 1835), 54; William Buckland, *Vindiciae Geologicae: Or the Connexion of Geology with Religion Explained* (Oxford: Oxford University Press, 1820), 12.

12. Michael Ruse, *The Darwinian Revolution: Science Red in Tooth and Claw* (Chicago: University of Chicago Press, 1979), 73.

13. On Malthus's concern regarding infant mortality among the poor, see Harriet Martineau, *Harriet Martineau's Autobiography* (London: Smith, Elder & Co., 1877), 159.

14. Patricia James, *Population Malthus: His Life and Times* (London: Routledge, 2013), 132.

15. James, *Population Malthus*, 43, 46, 164.

16. Thomas Malthus, *First Essay on Population 1798*, with notes by James Bonar (London: MacMillan and Co., 1926), iv; Robert Young, "Malthus and the Evolutionists: The Common Context of Biological and Social Theory," *Past and Present* 43 (1969): 132.

17. James, *Population Malthus*, 117.

18. Paley, *Natural Theology*, 540.

19. Malthus, *An Essay on the Principle of Population*, vol. 2 (London: John Murray, 1826), 267.

20. Malthus, *First Essay on Population 1798*, 363.

21. Malthus, *An Essay on the Principle of Population*, 287; Paley, *Natural Theology*, 540–41.

22. See Dennis Hodgson, "Malthus's Essay on Population and the American Debate over Slavery," *Comparative Studies in Society and History* 51, no. 4 (2009): 742–70.

23. Hodgson, "Malthus's Essay on Population and the American Debate over Slavery," 762.

24. Andrew Combe, *Treatise on the Physiological and Moral Management of Infancy* (Philadelphia: Carey & Hart, 1840), 11.

25. Combe, *Treatise on the Physiological and Moral Management of Infancy*, 25.

26. Anonymous, "Dr. Andrew Combe," *The Christian Reformer* 7, no. 74 (1851): 65–82; Combe, *Treatise on the Physiological and Moral Management of Infancy*, 29.

27. Combe, *Treatise on the Physiological and Moral Management of Infancy*, 18.

28. Ibid., 49.

29. George Combe, *The Life and Correspondence of Andrew Combe* (Edinburgh: MacLachlan and Stuart, 1850), 479–80.

30. Darwin, *The Life and Letters of Charles Darwin*, 36.

31. Ibid., 32.

32. Ibid., 45. For an analysis of skepticism regarding the sincerity of Darwin's religiosity, see David Kohn, "Darwin's Ambiguity: The Secularization of Biological Meaning," *The British Journal for the History of Science* 22, no. 2 (1989): 215–39.

33. John Pearson, *An Exposition of the Creed* (Oxford: Clarendon Press, 1797), 33, 53, 70, 72.

34. Darwin, *The Life and Letters of Charles Darwin*, 47.

35. Leonard Jenyns, *Memoir of the Rev. John Stevens Henslow* (London: John Van Voorst, 1862), 132.

36. Darwin, *The Life and Letters of Charles Darwin*, 45.

37. Jenyns, *Memoir of the Rev. John Stevens Henslow*, 136.

38. John Stevens Henslow, *A Sermon on the First and Second Resurrection* (Cambridge, UK: James Hodson, 1829), vii; Jenyns, *Memoir of the Rev. John Stevens Henslow*, 137.

39. Darwin, *The Life and Letters of Charles Darwin*, 53.

40. E. Harold Browne, *An Exposition of the Thirty-Nine Articles*, 6th ed. (London: Longman, Green, Longman, Roberts & Green, 1864), 13.

41. Jenyns, *Memoir of the Rev. John Stevens Henslow*, 131, and inset between pages 100 and 101.

42. Jenyns, *Memoir of the Rev. John Stevens Henslow*, 153–54, 176.

43. See Ronald L. Numbers, *The Creationists: From Scientific Creationism to Intelligent Design* (Cambridge, MA: Harvard University Press, 2006).

44. Jenyns, *Memoir of the Rev. John Stevens Henslow*, 218.

45. Ray, *The Wisdom of God*, 86.

46. Thomas Jefferson, *Notes on the State of Virginia* (Boston: Lilly and Wait, 1832), 83.

47. William Paley, *Natural Theology*, 515.

48. See Mark V. Barrow, *Nature's Ghosts: Confronting Extinction from the Age of Jefferson to the Age of Ecology* (Chicago: University of Chicago Press, 2009).

49. Alfred Tennyson, *In Memoriam*, 5th ed. (London: Edward Moxon, 1851), 80.

50. Benjamin Franklin, "The Death of Infants," *Pennsylvania Gazette*, 1734.

51. See Richard W. Burkhardt, "The Inspiration of Lamarck's Belief in Evolution," *Journal of the History of Biology* 5, no. 2 (1972): 413–38.

52. Adam Sedgwick, "President's Address," *Proceedings of the Geological Society of London* 1, no. 15 (1830): 207.

53. For Sedgwick's views on science and religion, see Browne, *Charles Darwin: Voyaging*, 137; for his speeches to coal miners, see Desmond, *The Politics of Evolution*, 173.

54. Lauren Cameron, "Mary Shelley's Malthusian Objections in *The Last Man*," *Nineteenth-Century Literature* 67, no. 2 (2012): 185, 187–88.

55. Brooke, *Science and Religion*, 194.

56. William Chilton, "Theory of Regular Gradation," *The Movement, Anti-persecution Gazette, and Register of Progress: A Weekly Journal of Republic Politics, Anti-Theology, & Utilitarian Morals* 1 (1845): 413. For Sedgwick's speech and Chilton's response, see Desmond and Moore, *Darwin*, 320. For John Kidd, see his *On the Adaptation of External Nature to the Physical Condition of Man, Principally with Reference to the Supply of His*

Wants, and the Exercise of His Intellectual Faculties (Philadelphia: Carey, Lea & Blanchard, 1836), 152. For Southwood Smith, see his *Illustrations of the Divine Government*, 1st American ed. (Boston: Benj. B Mussey, 1831), 28–29.

57. Charles Darwin to Caroline Darwin, April 25–26, 1832, DCP, "Letter no. 166."

58. Josiah Wedgwood II to R. W. Darwin, August 31, 1831, DCP, "Letter no. 109."

CHAPTER FOUR

1. C. R. Darwin, *1838–1839, Notebook N* [Metaphysics and Expression], CUL-DAR126, ed. Paul Barrett and John van Wyhe (Darwin Online), http://darwin-online.org.uk/), 36.

2. Darwin, *The Life and Letters of Charles Darwin*, 308.

3. Randal D. Keynes, ed., *Charles Darwin's Beagle Diary* (Cambridge, UK: Cambridge University Press, 1988), 403; Charles Darwin, *Journal of Researches into the Geology and Natural History of the Various Countries Visited by H.M.S. Beagle, under the Command of Captain Fitzroy, R.N. from 1832 to 1836*, 1st ed. (London: Henry Colburn, 1839), 605.

4. Paley, *Natural Theology*, 506; Charles Lyell, *Principles of Geology: Being an Inquiry How Far the Former Changes of the Earth's Surface are Referable to Causes now in Operation*, vol. 2 (London: John Murray, 1833), 58.

5. Browne, *Charles Darwin: Voyaging*, 190.

6. Lyell, *Principles of Geology*, vol. 2, 8.

7. Brooke, *Science and Religion*, 251–53.

8. Lyell, *Principles of Geology*, vol. 2, 15; Browne, *Charles Darwin: Voyaging*, 187.

9. On the Darwin family's relation to abolitionism, see Adrian Desmond and James Moore, *Darwin's Sacred Cause: How a Hatred of Slavery Shaped Darwin's Views on Human Evolution* (New York: Houghton Mifflin Harcourt, 2009).

10. Susan Darwin to Charles Darwin, October 15, 1833, DCP, "Letter no. 219."

11. J. B. Innes to Charles Darwin, May 26, 1871, DCP, "Letter no. 7768"; Darwin to J. B. Innes, May 29, 1871, DCP, "Letter no. 7776."

12. Charles Darwin to Catherine Darwin, May 22–July 14, 1833, DCP, "Letter no. 206."

13. Darwin, *Journal of Researches*, 2nd ed. (London: John Murray, 1845), 499.

14. Desmond and Moore, *Darwin's Sacred Cause*, 92.

15. Jefferson, *Notes on the State of Virginia*, 145; Heather Andrea Williams, *Help Me to Find My People: The African American Search for Family Lost in Slavery* (Chapel Hill: University of North Carolina Press, 2012), 97.

16. See Robin Bernstein, *Racial Innocence: Performing American Childhood from Slavery to Civil Rights* (New York: New York University Press, 2011), 50–51.

17. Wheatley, *Memoir and Poems*, 75.

18. Preface to Mary Prince's *The History of Mary Prince, a West Indian Slave. Related by Herself*, 3rd ed. (London: F. Westley and A. H. Davis, 1831), 4.

19. Darwin, *Journal of Researches*, 2nd ed., 499.

20. Ibid., 500.

21. Browne, *Charles Darwin: Voyaging*, 198.

22. Darwin, *Journal of Researches*, 2nd ed., 499–500.

23. Keynes, *Charles Darwin's* Beagle *Diary*, 45; Darwin, *Journal of Researches*, 2nd ed., 500.

24. Jemmy's given name was Orundellico. Captain Fitzroy called him Jemmy Button because he traded him for a mother of pearl button. The name Jemmy is used here because, according to the few records we have by travelers, Orundellico insisted on using his new name for the rest of his life. See the account in https://www.darwinproject.ac.uk /orundellico-jemmy-button; Darwin, *Journal of Researches*, 2nd ed., 207.

25. Browne, *Charles Darwin: Voyaging*, 147.

26. Robert Fitzroy, *Narrative of the Surveying Voyages of His Majesty's Ships Adventure and Beagle, Between the Years 1826–36, Describing Their Examination of the Southern Shores of South America, and the Beagle Circumnavigation of the Globe: In Three Volumes. Proceedings of the second expedition, 1831–1836, under the command of Captain Robert Fitz-Roy*, vol. 2 (London: Henry Colburn, 1839), 120–21.

27. Darwin, *Journal of Researches*, 2nd ed., 207.

28. Darwin, *Journal of Researches*, 2nd ed., 207, 214–15; see also Matthew Day, "Godless Savages and Superstitious Dogs: Charles Darwin, Imperial Ethnography, and the Problem of Human Uniqueness," *Journal of the History of Ideas* 69, no. 1 (2008): 49–70.

29. Darwin, *Journal of Researches*, 2nd ed., 221–22.

30. Charles Darwin to Caroline Darwin, March 30–April 12, 1833, DCP, "Letter no. 203."

31. Darwin, *Journal of Researches*, 2nd ed., 207–8, 213.

32. Browne, *Charles Darwin: Voyaging*, 246.

33. Charles Darwin to Catherine Darwin, April 6, 1834, DCP, "Letter no. 242."

34. Darwin, *Journal of Researches*, 1st ed., 236; Darwin, *Journal of Researches*, 2nd ed., 216.

35. Browne, *Charles Darwin: Voyaging*, 244–47.

36. Darwin, *Journal of Researches*, 2nd ed., 229; Charles Darwin to B. J. Sullivan, April 22, 1878, DCP, "Letter no. 11481."

37. For a more detailed account of these puzzles, see Browne, *Charles Darwin: Voyaging*, chapters 13–14.

38. Charles Darwin to Otto Zacharias, February 24, 1877, DCP, "Letter no. 10863."

39. Browne, *Charles Darwin: Voyaging*, 308–9.

40. Charles Darwin to Caroline Darwin, February 14, 1836, DCP, "Letter no. 298."

41. Browne, *Charles Darwin: Voyaging*, 331.

42. See Zoë Laidlaw, "Heathens, Slaves and Aborigines: Thomas Hodgkin's Critique of Missions and Anti-Slavery," *History Workshop Journal* 64, no. 1 (2007): 133–61.

43. Darwin, *Journal of Researches*, 1st ed., 520.

44. Ibid., 525.

45. Ibid., 520.

46. Ibid., 120–22.

47. Piers J. Hale, *Political Descent: Malthus, Mutualism, and the Politics of Evolution in Victorian England* (Chicago: University of Chicago Press, 2014), 95.

48. For historical work correcting the myth that Darwin discovered natural selection while observing finches on the Galapagos Islands, see Frank Sulloway, "Darwin and His

Finches: The Evolution of a Legend," *Journal of the History of Biology* 15, no. 1 (1982): 1–53.

49. C. R. Darwin, *1837–1838, Notebook B:* [Transmutation of Species], CUL-DARW121, ed. John van Wyhe (*Darwin Online*), http://darwin-online.org.uk), 44.

50. Desmond and Moore, *Darwin*, xvii.

51. Ray, *The Wisdom of God*, 94–95.

52. Darwin, *1837–1838, Notebook B*, 74.

53. Gavin de Beer, ed., *Darwin's Notebooks on Transmutation of Species Part II. Second Notebook (February to July 1838)* (London, 1960), 91.

54. de Beer, *Darwin's Notebooks*, 91.

55. Quoted in Desmond and Moore, *Darwin's Sacred Cause*, 115, from *Darwin's Notebooks*.

56. Charles Darwin, *On the Origin of Species by Means of Natural Selection, or the Preservation of Favoured Races in the Struggle for Life* (London: John Murray, 1859), 3.

CHAPTER FIVE

1. Emma Darwin to Charles Darwin, April 24, 1851, DCP, "Letter no. 1414."

2. Desmond, *The Politics of Evolution*, 125–26.

3. Cobbett is quoted in Barry G. Gale, "Darwin and the Concept of a Struggle for Existence: A Study in the Extra-Scientific Origins of Scientific Ideas," *Isis* 63, no. 3 (1972): 338.

4. Charles Darwin to Caroline Darwin, December 7, 1836, DCP, "Letter no. 325."

5. Martineau, *Autobiography*, 309.

6. Malthus, *Essay on Population 1798*, 158.

7. Peter Bowler, "Malthus, Darwin, and the Concept of Struggle," *Journal of the History of Ideas* 37, no. 4 (1976): 642.

8. Darwin, *The Life and Letters of Charles Darwin*, 83.

9. On influential Christian ministers and naturalists' support for Darwin's ideas, see Moore, *The Post-Darwinian Controversies*; and Livingstone, *Darwin's Forgotten Defenders*.

10. Darwin, *The Life and Letters of Charles Darwin*, 307–9.

11. Charles Darwin, "This Is the Question Marry Not Marry" [Memorandum on marriage], 1838, CUL-DAR210.8.2, ed. John van Wyhe, Darwin Online, http://darwin-online.org.uk/.

12. Darwin, *The Life and Letters of Charles Darwin*, 13; see also Evelleen Richards, "Darwin and the Descent of Woman," in *The Wider Domain of Evolutionary Thought*, ed. David Oldroyd and Ian Langham (Dordrecht: Springer, 1983), and *Darwin and the Making of Sexual Selection* (Chicago: University of Chicago Press, 2017).

13. Emma Darwin to Charles Darwin, November 21–22, 1838, DCP, "Letter no. 441."

14. Emma Darwin to Charles Darwin, February 1839, DCP, "Letter no. 471."

15. Henrietta E. Litchfield, *Emma Darwin: Wife of Charles Darwin*, vol. 2 (Cambridge, UK: Cambridge University Press, 1904), 187; Browne, *Charles Darwin: Voyaging*, 173, 397.

16. Darwin, *The Life and Letters of Charles Darwin*, 308.

17. Litchfield, *Emma Darwin*, 50.

18. Emma Darwin to Charles Darwin, June 1861, DCP, "Letter no. 3169."

19. Anonymous, Review of "The Child of the Islands. A Poem. By the Hon. Mrs. Norton," *Edinburgh Review* 82 (1845): 92.

20. Elizabeth Barrett Browning, *The Poetical Works of Elizabeth Barrett Browning* (New York: Macmillan, 1897), 249.

21. George Jacob Holyoake, *The History of the Last Trial by Jury for Atheism in England* (London: James Watson, 1850), 5.

22. Holyoake, *The History of the Last Trial by Jury*, 75.

23. Ibid., 84–85; George Jacob Holyoake, *Paley Refuted in His Own Words* (London: Hetherington, 1847), iii; Adrian Desmond, *Huxley: From Devil's Disciple to Evolution's High Priest* (Reading, MA: Addison-Wesley, 1994), 160.

24. Charles Darwin to John Stevens Henslow, January 17, 1850, DCP, "Letter no. 1293."

25. Charles Darwin to Joseph Dalton Hooker, January 11, 1844, DCP, "Letter no. 729."

26. Charles Darwin to Emma Darwin, July 5, 1844, DCP, "Letter no. 761."

27. R. C. Lehmann, "Memories of Half a Century," *Chambers's Journal* 80 (1903): 50.

28. James A. Secord, *A Victorian Sensation: The Extraordinary Publication, Reception, and Secret Authorship of* Vestiges of the Natural History of Creation (Chicago: University of Chicago Press, 2000), 80; Secord, "Behind the Veil: Robert Chambers and Vestiges," in *History, Humanity and Evolution: Essays for John C. Greene*, ed. James Moore (Cambridge, UK: Cambridge University Press, 1989), 194, footnote 71.

29. Robert Chambers, *Vestiges of the Natural History of Creation* (London: John Churchill, 1844), 152–56, 233; Anonymous, "Children," *Chambers's Edinburgh Journal* 1, no. 34 (1832): 266.

30. Chambers, *Vestiges*, 156, 203, 234.

31. Ibid., 198; William Chambers, *Memoir of Robert Chambers, with Autobiographic Reminiscences of William Chambers* (Edinburgh and London: W&R Chambers, 1872), 47.

32. Chambers, *Vestiges*, 362–64.

33. Brooke, *Science and Religion: Historical Perspectives*, 153.

34. Chambers, *Vestiges*, 368, 370–73.

35. Adam Sedgwick (published anonymously), "Review of *Vestiges of the Natural History of Creation*," *The Edinburgh Review* 82 (1845): 2, 64.

36. Hugh Miller, *Foot-Prints of the Creator: or, The Asterolepis of Stromness* (London: Johnstone and Hunter, 1849), 262–63.

37. Darwin, *The Life and Letters of Charles Darwin*, 333. On the reviews of *Vestiges*, see Secord, *A Victorian Sensation*.

38. Alfred Russel Wallace, *The Wonderful Century: Its Successes and Its Failures* (London: Swann Sonnenschein, 1898), 137; Chambers, *Vestiges*, 219; Anonymous, "Review of *Vestiges of the Natural History of Creation*," *The Lancet* 1, no. 2 (1845): 178.

39. Secord, *A Victorian Sensation*, 203–7.

40. Robert Chambers, *Explanations: A Sequel to "Vestiges of the Natural History of Creation"* (New York: Wiley & Putnam, 1846), 129–30.

41. Darwin to Hooker, September 10, 1845, DCP, "Letter no. 915."

42. Jean Beagle Ristaino and Donald H. Pfister, "'What a Painfully Interesting Subject': Charles Darwin's Studies of Potato Late Blight," *BioScience* 66, no. 12 (2016): 1035–45.

43. Jenyns, *Memoir of the Rev. John Stevens Henslow*, 204.

44. Browne, *Charles Darwin: Voyaging*, 474.

45. Litchfield, *Emma Darwin*, 143.

46. Ibid., 146.

47. William Owen to Charles Darwin, March 26, 1843, DCP, "Letter no. 666"; Litchfield, *Emma Darwin*, 147.

48. Litchfield, *Emma Darwin*, 147–49.

49. See Edward B. Aveling, *The Religious Views of Charles Darwin* (London: Freethought Publishing Company, 1883). For an analysis of (and debate regarding) the role of Annie's death in Charles's changing beliefs, see Randal Keynes, *Darwin, His Daughter, and Human Evolution* (New York: Penguin, 2002); James R. Moore, "Of Love and Death: Why Darwin 'Gave Up Christianity,'" in *History, Humanity, and Evolution*, 195–229; and John Van Wyhe and Mark J. Pallen, "The 'Annie Hypothesis': Did the Death of His Daughter Cause Darwin to 'Give Up Christianity'?" *Centaurus* 54, no. 2 (2012): 105–23. For additional studies of Darwin's changing religious beliefs, see Maurice Mandelbaum, "Darwin's Religious Views," *Journal of the History of Ideas* 19, no. 3 (1958): 363–78; and Frank Burch Brown, "The Evolution of Darwin's Theism," *Journal of the History of Biology* 19, no. 1 (1986): 1–45.

50. Nora Barlow, ed., *The Autobiography of Charles Darwin, 1809–1882, with original omissions restored* (London: Collins, 1958), 90.

51. Martineau, *Autobiography*, 28, 484–85.

52. Charles Darwin to Fanny Mackintosh Wedgwood, April 24, 1851, DCP, "Letter no. 1413"; Darwin, *Notebook N*, 12.

53. Charles Darwin to Asa Gray, May 22, 1860, DCP, "Letter no. 2814"; Darwin to Mary Everest Boole, December 14, 1866, DCP, "Letter no. 5307."

54. These lines were restored in a 1958 edition of the autobiography. For more on this fascinating story, see James R. Moore's account in *The Darwin Legend* (Grand Rapids, MI: Baker Books, 1994) of the family's reasons for publishing.

55. Barlow, *The Autobiography of Charles Darwin*, 87. Emma's shorthand has been changed into complete words.

56. Pearson, *An Exposition of the Creed*, 366.

57. Howard R. Murphy, "The Ethical Revolt against Christian Orthodoxy in Early Victorian England," *The American Historical Review* 60, no. 4 (1955): 800–801.

58. Darwin, *The Life and Letters of Charles Darwin*, 311.

59. Darwin to Hooker, May 9, 1856, DCP, "Letter no. 1870"; Darwin to J. D. Dana, September 29, 1856, DCP, "Letter no. 1964."

60. Leonard G. Wilson, ed., *Sir Charles Lyell's Scientific Journals on the Species Question* (New Haven, CT: Yale University Press, 1970), 264, 289.

61. Ibid.

62. Darwin to Hooker, May 9 and 11, 1856, DCP, "Letter no. 1870 and 1874."

63. Charles Darwin to Charles Lyell, June 18, 1858, DCP, "Letter no. 2285."

64. Alfred Russel Wallace, *My Life: A Record of Events and Opinions*, vol. 1 (New York: Dodd, Mead and Company, 1905), 226–28.

65. This is the form in which the question was put by the ancient Greek philosopher Epicurus.

66. Wallace, *My Life*, 87.

67. Ibid., 88, 226–28.

68. Quoted in Brooke, *Science and Religion*, 219; John R. Durant, "Scientific Naturalism and Social Reform in the Thought of Alfred Russel Wallace," *British Journal for the History of Science* 12, no. 1 (1979): 38–39; Wallace, *My Life*, 29.

69. Peter Raby, *Alfred Russel Wallace: A Life* (London: Chatto and Windus, 2001), 173; Henry Walter Bates, *The Naturalist on the River Amazons* (London: John Murray, 1863), iii; Wallace, *My Life*, 255–56.

70. Alfred Russel Wallace, *A Narrative of Travels on the Amazon and Rio Negro, with an account of the native tribes, and observations of the climate, geology, and natural history of the Amazon valley* (London: Reeve and Co., 1853), 401.

71. Wallace, *My Life*, 361.

72. Wallace, *My Life*, 362.

73. Charles Lyell and Joseph Dalton Hooker to the Linnean Society, June 30, 1858, DCP, "Letter no. 2299."

74. Janet Browne, *Charles Darwin: The Power of Place* (Princeton, NJ: Princeton University Press, 2002), 34.

75. Caspar Morris, *Lectures on Scarlet Fever* (Philadelphia: Lindsay & Blakiston, 1851), 1.

76. Darwin to Hooker, June 29, 1858, and June 29, 1858 (evening), DCP, "Letters no. 2297 and 2298."

77. Charles Darwin and Alfred Wallace, "On the Tendency of Species to Form Varieties; and on the Perpetuation of Varieties and Species by Natural Means of Selection," *Zoological Journal of the Linnean Society* 3, no. 9 (1858): 45.

78. Charles Darwin, *On the Origin of Species by Means of Natural Selection, or the Preservation of Favoured Races in the Struggle for Life* (London: John Murray, 1859), 67, 79.

79. Darwin and Wallace, "On the Tendency of Species to Form Varieties," 48–49.

80. Browne, *Charles Darwin: Voyaging*, 542.

Chapter Six

1. Barlow, *The Autobiography of Charles Darwin*, 93.

2. Darwin to Hooker, April 23, 1861, DCP. "Letter no. 3098"; Charles Darwin, *On the Origin of Species*, 3.

3. Darwin, *On the Origin of Species*, 5.

4. Ibid., 62.

5. Ibid., 471.

6. Ibid., 244.

7. Ibid., 472.

8. Ibid.

9. Ibid.

10. Benjamin Franklin, "The Death of Infants," *Pennsylvania Gazette*, 1734.

11. Darwin, *On the Origin of Species*, 472.

12. Benjamin Waterhouse, *An Essay Concerning Tussis Convulsiva, or, Whooping Cough. With Observations on the Diseases of Children* (Boston: Munroe and Francis, 1822), 130.

13. Darwin, *On the Origin of Species*, 490.

14. Barlow, *The Autobiography of Charles Darwin*, 90.

15. Jenyns, *Memoir of the Rev.*, 213.

16. Adam Sedgwick to Charles Darwin, November 24, 1859, DCP, "Letter no. 2548."

17. Darwin to Lyell, March 24–April 3, 1860, DCP, "Letter no. 2734."

18. See, for example, John Gulick, *Evolutionist and Missionary, John Thomas Gulick, Portrayed through Documents and Discussions* (Chicago: University of Chicago Press, 1932), 165. For a detailed survey and analysis, see Moore, *The Post-Darwinian Controversies*.

19. Asa Gray, *Darwiniana: Essays and Reviews Pertaining to Darwinism* (New York: D. Appleton and Company, 1877), 378.

20. Asa Gray, *Natural Selection not Inconsistent with Natural Theology: A Free Examination of Darwin's Treatise on the Origin of Species, and of its American Reviewers*, reprinted from *The Atlantic Monthly*, July, August, and October 1860 (London: Trübner and Co., 1861), 45. For Jane's testimony regarding Gray's belief, see Moore, *The Post-Darwinian Controversies*, 271.

21. Darwin to Gray, May 22, 1860, DCP, "Letter no. 2814."

22. Asa Gray, "Review of *Vestiges of the Natural History of Creation*, by Robert Chambers," *The North American Review* 60, no. 127 (1845): 475.

23. Gray to Darwin, November 24, 1862, DCP, "Letter no. 3823."

24. Samuel Wilberforce, "Review of *On the Origin of Species*, by Charles Darwin," *Quarterly Review* (1860): 121.

25. Wilberforce, "Review of *On the Origin of Species*," 122, 136.

26. Ibid., 137.

27. Darwin to Lubbock, November 22, 1859, DCP, "Letter no. 2532."

28. Leonard Huxley, ed., *Life and Letters of Thomas Henry Huxley*, vol. 1 (London: Macmillan and Co., 1900), 170.

29. Thomas Henry Huxley to Darwin, November 23, 1859, DCP, "Letter no. 2544." The Gospel reference is from a letter from Darwin to Huxley from August 8, 1860, in Darwin, The *Life and Letters of Charles Darwin*, 331.

30. Desmond, *Huxley*, 3.

31. Huxley, *Life and Letters of Thomas Henry Huxley*, 16–17.

32. Huxley, *Life and Letters of Thomas Henry Huxley*, 17; Desmond, *Huxley*, 10.

33. Desmond, *Huxley*, 138.

34. T. H. Huxley, "Review VII. *Vestiges of the Natural History of Creation*, Tenth Edition, London, 1853," *The British and Foreign Medico-Chirurgical Review* 13, no. 13 (1854): 425–26; Huxley, *Life and Letters of Thomas Henry Huxley*, 187–202.

35. Darwin to Huxley, July 20, 1860, DCP, "Letter no. 2873"; Darwin to Hooker, July 20, 1860, DCP, "Letter no. 2875."

36. Wilberforce, "Review of *On the Origin of Species*," 135.

37. Quoted in J. Vernon Jensen, "Return to the Wilberforce–Huxley Debate," *The British Journal for the History of Science* 21, no. 2 (1988): 168.

38. On the Huxley-Wilberforce debate as both legend and historical reality, see John Randolph Lucas, "Wilberforce and Huxley: A Legendary Encounter," *The Historical Journal* 22, no. 2 (1979): 313–30; and Jensen, "Return to the Wilberforce–Huxley Debate."

39. Leslie Stephen is quoted in Martin Huxley Cooke, *The Evolution of Nettie Huxley: 1825–1914* (Chichester, UK: Phillimore, 2008), 106; Huxley, *Life and Letters of Thomas Henry Huxley*, 158.

40. Cooke, *The Evolution of Nettie Huxley*, 16, 36, 65.

41. Ibid., 64–65; Baden Powell, "On the Study of the Evidences of Christianity," in *Essays and Reviews*, 2nd ed. (London: John W. Parker & Son, 1860), 139.

42. Darwin to Huxley, September 10, 1860, DCP, "Letter no. 2909."

43. Quoted in Desmond, *Huxley*, 286.

44. Huxley, *Life and Letters of Thomas Henry Huxley*, 152.

45. Quoted in Desmond, *Huxley*, 291.

46. Darwin to Huxley, September 18, 1860, DCP, "Letter no. 2920B."

47. Darwin to Huxley, February 22, 1861, DCP, "Letter no. 3066"; Cooke, *The Evolution of Nettie Huxley*, 74; Browne, *Darwin: The Power of Place*, 152.

48. Huxley, *Life and Letters of Thomas Henry Huxley*, 222.

49. Huxley, *Life and Letters of Thomas Henry Huxley*, 213.

50. Huxley's letter is reprinted in Huxley, *Life and Letters of Thomas Henry Huxley*, 217–22.

51. See Matthew Stanley, *Huxley's Church and Maxwell's Demon: From Theistic Science to Naturalistic Science* (Chicago: University of Chicago Press, 2015); Ronald L. Numbers, *Science and Christianity in Pulpit and Pew* (Oxford: Oxford University Press, 2007); and Geoffrey Cantor, "Science, Providence, and Progress at the Great Exhibition," *Isis* 103, no. 3 (2012): 439–59. For more on Huxley's campaign as arising from an alternative explanation of suffering, see James R. Moore, "Theodicy and Society: The Crisis of the Intelligentsia," in *Victorian Faith in Crisis*, ed. Richard J. Helmstadter and Bernard Lightman (London: Palgrave Macmillan, 1990), 153–86.

52. Wilberforce, "Review of *On the Origin of Species*," 135.

53. Darwin, *On the Origin of Species*, 488.

54. Huxley, *Life and Letters of Thomas Henry Huxley*, 320; Leonard Huxley, ed., *Life and Letters of Thomas Henry Huxley*, vol. 3 (London: Macmillan Company, 1913), 97–98. On Huxley and the history of agnosticism, see Bernard Lightman, *The Origins of Agnosticism: Victorian Unbelief and the Limits of Knowledge* (Baltimore: Johns Hopkins University Press, 1987).

55. Quoted in Desmond, *Huxley*, 275.

56. On shifts in child mortality rates, see Robert Millward and Frances Bell, "Infant Mortality in Victorian Britain: The Mother as Medium," *Economic History Review* 54, no. 4 (2001): 701.

Chapter Seven

1. See Desmond and Moore, *Darwin's Sacred Cause*, 291–93.

2. Darwin, *The Life and Letters of Charles Darwin*, 312–13.

3. Emma Darwin to Frances Power Cobbe, February 25, 1871, DCP, "Letter no. 7516F."

4. On Darwin's strategy with respect to race in *The Descent*, see Joseph L. Yannielli, "A Yahgan for the Killing: Murder, Memory and Charles Darwin," *The British Journal for the History of Science* 46, no. 3 (2013): 415–43; and Suman Seth, "Darwin and the Ethnologists: Liberal Racialism and the Geological Analogy," *Historical Studies in the Natural Sciences* 46, no. 4 (2016): 490–527.

5. Darwin to Hooker, March 29, 1863, DCP, "Letter no. 4065." *Pentateuchal* refers to the first five books of the Old Testament, including Genesis.

6. Leonard Huxley, ed., *Life and Letters of Sir Joseph Dalton Hooker* (London: John Murray, 1918), 106.

7. Hooker to Darwin, September 28, 1863, DCP, "Letter no. 4309"; Hooker to Darwin, October 1, 1863, DCP, "Letter no. 4317."

8. Darwin to Hooker, October 4, 1863, DCP, "Letter no. 4318"; Hooker to Darwin, October 23, 1863, DCP, "Letter no. 4321."

9. Huxley, *Life and Letters of Sir Joseph Dalton Hooker*, 65; Charles Larcom Graves, *The Life & Letters of Sir George Grove* (London: Macmillan and Co., 1903), 94; Hooker to Darwin, September 16, 1864, DCP, "Letter no. 4614."

10. Huxley, *Life and Letters of Sir Joseph Dalton Hooker*, 64.

11. Paley, *Natural Theology* (1809 ed.), 473.

12. Charles Lyell, *The Geological Evidences of the Antiquity of Man, with Remarks on Theories of the Origin of Species by Variation* (London: John Murray, 1863), 503.

13. Lyell, *The Geological Evidences of the Antiquity of Man*, 504.

14. Darwin to Gray, February 23, 1863, DCP, "Letter no. 4006"; Darwin to Lyell, March 6, 1863, DCP, "Letter no. 4028"; Lyell to Darwin, March 11, 1863, DCP, "Letter no. 4035."

15. Darwin to Hooker, February 24, 1863, DCP, "Letter no. 4009."

16. Wallace, *My Life*, 372–74. The italics are Wallace's.

17. James Marchant, ed., *Alfred Russel Wallace: Letters and Reminiscences*, vols. 1 and 2 (London: Cassell and Company, 1916), 67.

18. Alfred Russel Wallace, "The Origin of Human Races and the Antiquity of Man Deduced From the Theory of 'Natural Selection,'" *Journal of the Anthropological Society of London* 2 (1864): clxv.

19. Wallace, "The Origin of Human Races," clxxv, clxv.

20. Darwin to Wallace, May 28, 1864, DCP, "Letter no. 4510."

21. Wallace, *My Life*, 361.

22. Wallace, *A Defence of Modern Spiritualism* (Boston: Colby and Rich, 1874), 21, 35, 36; Wallace, *On Miracles and Modern Spiritualism* (London: James Burns, 1875), vii.

23. Malcolm Jay Kottler, "Alfred Russel Wallace, the Origin of Man, and Spiritualism," *Isis* 65, no. 2 (1974): 150; Wallace, "Sir Charles Lyell on Geological Climates and the Origin of Species," *Quarterly Review* 126, no. 252 (1869): 391–392, 394.

24. Darwin to Wallace, March 27, 1869, DCP, "Letter no. 6684"; Darwin to Wallace, April 14, 1869, DCP, "Letter no. 6706"; Marchant, *Alfred Russel Wallace*, 251.

25. Bernard Lightman, "Huxley and the Devonshire Commission," in *Victorian Scientific Naturalism*, ed. Gowan Dawson and Bernard Lightman (Chicago: University of Chicago Press, 2014), 101.

26. Darwin, *The Descent of Man*, 3; Amalie M. Kass and Edward Harold Kass, *Perfecting the World: The Life and Times of Dr. Thomas Hodgkin 1798–1866* (Boston: Harcourt Brace Jovanovich, 2008), 161.

27. Darwin, *The Descent of Man*, 25.

28. Ernst Haeckel, *The Evolution of Man: A Popular Study* (New York: G. P. Putnam's Sons, 1910), 72.

29. Robert Richards, *The Tragic Sense of Life: Ernst Haeckel and the Struggle over Evolutionary Thought* (Chicago: University of Chicago Press, 2008), 105; Charles Darwin to Ernst Haeckel, March 3, 1864, DCP, "Letter no. 4419"; Haeckel to Darwin, July 9, 1864, DCP, "Letter no. 4555"; Darwin, *The Descent of Man*, 142.

30. Darwin, *The Descent of Man*, 70.

31. Ibid., 5–6.

32. Ibid., 203–4.

33. Darwin, *On the Origin of Species*, 204; Darwin, *The Descent of Man*, 180. See also Day, "Godless Savages and Superstitious Dogs."

34. Darwin, *The Descent of Man*, 65, 178.

35. Darwin, *The Descent of Man*, 66. On Craft's riposte, see Desmond, *Darwin's Sacred Man*, 338–39.

36. Charles William Grant, *Our Blood Relations, or the Darwinian Theory* (London: Simpkin, Marshall & Co., 1872), 81–82.

37. Darwin, *Journal of Researches*, 1st ed., 525.

38. Darwin, *The Descent of Man*, 130.

39. This and subsequent paragraphs are from Darwin, *The Descent of Man*, 182–92.

40. Darwin, *The Descent of Man*, 125.

41. Charles Darwin to Charles Kingsley, February 6, 1862, DCP, "Letter no. 3439"; Darwin, *The Descent of Man*, 183; Desmond and Moore, *Darwin's Sacred Cause*, 146–47.

42. Kingsley to Darwin, January 30, 1862, DCP, "Letter no. 3426"; Darwin to Kingsley, February 6, 1862, DCP, "Letter no. 3439."

43. Darwin, *The Descent of Man*, 138–39.

44. Unknown correspondent to Darwin, June 13, 1877, DCP, "Letter no. 10998"; Darwin, *The Descent of Man*, 138–39.

45. Darwin, *The Descent of Man*, 142.

46. Charles Darwin to William Graham, July 3, 1881, DCP, "Letter no. 13230."

47. Te Rangi Hiroa (P. H. Buck), "The Passing of the Maori," *Transactions and Proceedings of the New Zealand Institute* 55 (1924): 363.

48. Darwin, *The Descent of Man*, 122–23.

Chapter Eight

1. J. D. Hague, "A Reminiscence of Mr. Darwin," *Harper's New Monthly Magazine* 69, issue 413 (1884): 760.

2. Diane B. Paul, "Darwin, Social Darwinism and Eugenics," in *The Cambridge Companion to Darwin*, ed. Jonathan Hodge and Gregory Radick (Cambridge, UK: Cambridge University Press, 2006), 236; Richard Hofstadter, *Social Darwinism in American Thought: 1860–1915* (Philadelphia: University of Pennsylvania Press, 1944), 74.

3. David W. Blight, *Frederick Douglass: Prophet of Freedom* (New York: Simon & Schuster, 2020), 319.

4. Blight, *Frederick Douglass*, 569–70.

5. Blight, *Frederick Douglass*, 115; Frederic May Holland, *Frederick Douglass: The Colored Orator* (New York: Funk and Wagnalls Company, 1895), 42.

6. James McCune Smith, "Colored Orphan's Asylum: Physicians Report," *The Colored American*, January 26, 1839. On Smith, see Anna Mae Duane, *Educated for Freedom: The Incredible Story of Two Fugitive Schoolboys Who Grew Up to Change a Nation* (New York: New York University Press, 2020); and John Stauffer, ed., *The Works of James McCune Smith: Black Intellectual and Abolitionist* (Oxford: Oxford University Press, 2006).

7. For mortality rates, see Richard H. Steckel, "A Dreadful Childhood: The Excess Mortality of American Slaves," *Social Science History* 10, no. 4 (1986): 427. For Nott's arguments, see Josiah C. Nott and George R. Gliddon, *Types of Mankind* (London: Trübner & Co., 1854).

8. Josiah C. Nott, "The Mulatto a Hybrid—Probable Extermination of the two races if Whites and Blacks are allowed to Marry," *The American Journal of the Medical Sciences* 5 (1843): 254. The italics are Nott's.

9. Frederick Douglass, *The Claims of the Negro, Ethnologically Considered* (Rochester, NY: Lee, Mann & Co, 1854), 8.

10. For the place of Darwin's work in Victorian debates over race, see Desmond and Moore, *Darwin's Sacred Cause;* and John C. Greene, "Darwin as a Social Evolutionist," *Journal of the History of Biology* 10, no. 1 (1977): 1–27. For abolitionist hopes that Darwin's work would serve the cause, see Randall Fuller, *The Book That Changed America: How Darwin's Theory of Evolution Ignited a Nation* (New York: Penguin, 2018).

11. Thomas Henry Huxley, *Science and Education: Essays* (New York: D. Appleton and Company, 1898), 66–68.

12. Desmond and Moore, *Darwin's Sacred Cause*, 318. On the statements in the US congressional investigations, see David S. Jones, *Rationalizing Epidemics: Meanings and Uses of American Indian Mortality since 1600* (Cambridge, MA: Harvard University Press, 2004), 138.

13. David Friedrich Strauss, *The Old Faith and the New: A Confession*, 2nd ed. (London: Asher & Co., 1873), 205; Brooke, *Science and Religion*, 271. On Douglass's bust of Strauss, see Holland, *Frederick Douglass*, 339.

14. Frederick Douglass, *The Essential Douglass: Selected Writings and Speeches* (Hackett Publishing, 2016), 286–300. For analysis, see D. H. Dilbeck, *Frederick Douglass: America's Prophet* (Chapel Hill: University of North Carolina Press, 2018), 148.

15. For Chickering, see Marshall Hall, *The Two-Fold Slavery of the United States* (London: Adam Scott, 1854), 19.

16. Douglass, *The Essential Douglass*, 289.

17. Albion Winegar Tourgée, *An Appeal to Caesar* (New York: Fords, Howard and Hulbert, 1884), 127–28; John S. Haller, "The Physician versus the Negro: Medical and Anthropological Concepts of Race in the Late Nineteenth Century," *Bulletin of the History of Medicine* 44, no. 2 (1970): 155.

18. Benjamin Waterhouse, *An Essay Concerning Tussis Convulsiva*, 20.

19. Annie Besant, *Annie Besant: An Autobiography*, 2nd ed. (London: T. Fisher Unwin, 1893. Originally published in 1885), 87.

20. Besant, *Annie Besant: An Autobiography*, 87, 89.

21. Ibid., 90–91.

22. Ibid., 97–98, 100–102.

23. Ibid., 92–93.

24. Ibid., 98.

25. Ibid., 129, 146.

26. Ibid., 152, 165–66.

27. James, *Population Malthus*, 61.

28. James, *Population Malthus*, 386–87; Darwin, *The Descent of Man*, 591–92. See also Marguerita Stephens, *White without Soap: Philanthropy, Caste and Exclusion in Colonial Victoria 1835–1888: A Political Economy of Race* (University of Manitoba Custom Book Centre, 2010), 102. For the history of tales of infanticide as a justification of colonialism, see Jeffrey Tobin, "Savages, the Poor and the Discourse of Hawaiian Infanticide," *The Journal of the Polynesian Society* 106, no. 1 (1997): 65–92.

29. Charles Bradlaugh and Annie Besant, "Publisher's Preface" to *Fruits of Philosophy: An Essay on the Population Question, by Charles Knowlton* (Rotterdam: Van der Hoeven & Buys, 1878), vi.

30. Desmond and Moore, *Darwin's Sacred Cause*, 627; Charles Darwin to Charles Bradlaugh, June 6, 1877, DCP, "Letter no. 10988."

31. Darwin, *The Descent of Man*, 618.

32. Mytheli Sreenivas, "Birth Control in the Shadow of Empire: The Trials of Annie Besant, 1877–1878," *Feminist Studies* 41, no. 3 (2015): 509–37.

33. Besant, *Annie Besant: An Autobiography*, 218, 220.

34. Besant, *Annie Besant: An Autobiography*, 163, 169, 244, 246.

35. Mahatma Gandhi, *A Guide to Health* (Madras: S. Ganesan, 1921), 81.

36. Quoted in Raby, *Alfred Russel Wallace*, 212.

37. Marchant, *Alfred Russel Wallace: Letters and Reminiscences*, vol. 1, 422.

38. Kottler, "Alfred Russel Wallace, the Origin of Man, and Spiritualism," 183; Wallace, *A Defence of Modern Spiritualism*, 16, 29.

39. Huxley, "Report on Spiritualism," *Daily News*, October 17, 1871; Kottler, "Alfred Russel Wallace, the Origin of Man, and Spiritualism," 182; Wallace, *A Defence of Modern Spiritualism*, 57.

40. Desmond, *Huxley*, 495; T. Jeffery Parker, "Professor Huxley: From the Point of View of a Disciple," *Natural Science: A Monthly Review of Scientific Progress* 8 (1896): 166.

41. Desmond, *Huxley*, 393. See also Nancy Tomes, *The Gospel of Germs: Men, Women, and the Microbe in American Life* (Cambridge, MA: Harvard University Press, 1999).

42. On Huxley, Wallace, and debates over vaccination, see chapter 5 of Rob Boddice's *The Science of Sympathy: Morality, Evolution, and Victorian Civilization* (Springfield: University of Illinois Press, 2016); Thomas P. Weber, "Alfred Russel Wallace and the Antivaccination Movement in Victorian England," *Emerging Infectious Disease* 16, no. 4 (2010): 668; and Martin Fichman and Jennifer E. Keelan, "Resister's Logic: The Anti-Vaccination Arguments of Alfred Russel Wallace and Their Role in the Debates over Compulsory Vaccination in England, 1870–1907," *Studies in History and Philosophy of Science Part C: Studies in History and Philosophy of Biological and Biomedical Sciences* 38, no. 3 (2007): 585–607.

43. Anonymous, "The Department of Sanitation, Food, and Public Health," *Medical Examiner* 3 (1878): 555; Arthur Newsholme, *Fifty Years in Public Health: A Personal Narrative with Comments* (London: George Allen & Unwin, 1935), 189.

44. Anonymous, "Diphtheria and Antitoxin at the Glasgow Fever Hospital," *The Lancet* 147, no. 3798 (1896): 1668.

45. Evelynn Hammonds, *Childhood's Deadly Scourge: The Campaign to Control Diphtheria in New York City, 1880–1930* (Baltimore, MD: Johns Hopkins University Press, 2002), 6–7.

46. Darwin, *The Descent of Man*, 159.

47. Darwin, *The Descent of Man*, 134. On Darwin's ambivalence regarding eugenics, see Paul, "Darwin, Social Darwinism and Eugenics."

48. Francis Galton, *Memories of My Life* (Yorkshire: Methuen, 1908), 323; Francis Galton, "Eugenics: Its Definition, Scope, and Aims," *The American Journal of Sociology* 10, no. 1 (1904): 5. On the history of eugenics in general, see Robert A. Nye, "The Rise and Fall of the Eugenics Empire: Recent Perspectives on the Impact of Biomedical Thought in Modern Society," *The Historical Journal* 36, no. 3 (1993): 687–70; Diane B. Paul, *Controlling Human Heredity: 1865 to the Present* (New York: Prometheus Books, 1995); and Alison Bashford and Philippa Levine, eds., *The Oxford Handbook of the History of Eugenics* (Oxford: Oxford University Press, 2010).

49. On Dean Inge's support for eugenics, see Christine Rosen, *Preaching Eugenics: Religious Leaders and the American Eugenics Movement* (Oxford: Oxford University Press, 2004), 133.

50. Marchant, *Alfred Russel Wallace: Letters and Reminiscences*, vol. 2, 246–47. For a study of the broad ideological appeal of eugenics, see Diane B. Paul, "Eugenics and the Left," *Journal of the History of Ideas* 45, no. 4 (1984): 567–90.

51. Boddice, *The Science of Sympathy*, 127; Frederick S. Crum, "The Effect of Infant Mortality on the After-Lifetime of Survivors," *American Child Hygiene Association: Transactions of the Tenth Annual Meeting*, 177–94 (Albany, NY: J. B. Lyon, 1920), 179; Caleb Williams Saleeby, *Parenthood and Race Culture: An Outline of Eugenics* (New York: Moffat, Yard and Company, 1915; first printing 1909), ix and 24–26. The italics are Saleeby's.

52. Saleeby, *Parenthood and Race Culture*, vii, 32–33.

53. Thomas Henry Huxley, *Evolution and Ethics, and Other Essays* (New York: D. Appleton and Co., 1894), 22–23, 39.

54. Charles Darwin to Francis Galton, January 4, 1873, DCP, "Letter no. 8724"; Darwin, *The Descent of Man*, 617–18.

55. Owen to Darwin, March 26, 1843, DCP, "Letter no. 666."

56. Robert Chambers, *Explanations: A Sequel to "Vestiges of the Natural History of Creation"* (New York: Wiley & Putnam, 1846), 129–30.

Conclusion

1. See David Paul Crook, *Darwinism, War and History: The Debate over the Biology of War from the "Origin of Species" to the First World War* (Cambridge, UK: Cambridge University Press, 1994). For scholars debating the relationship between Darwin's work and Nazi ideology, see Richard Weikart, *From Darwin to Hitler: Evolutionary Ethics, Eugenics and Racism in Germany* (Dordrecht: Springer, 2016), and Robert J. Richards, *Was Hitler a Darwinian?: Disputed Questions in the History of Evolutionary Theory* (Chicago: University of Chicago Press, 2013).

2. See Michael Ruse, *The Problem of War: Darwinism, Christianity, and the Battle to Understand Human Conflict* (Oxford: Oxford University Press, 2019).

3. William Jennings Bryan, *In His Image* (Chicago: Fleming H. Revell Co., 1922), 111.

4. Bryan's 1904 speech is quoted in Edward J. Larson, *Summer for the Gods: The Scopes Trial and America's Continuing Debate over Science and Religion* (New York: Basic Books, 1997), 39.

5. See Larson, *Summer for the Gods*. On the rise of "scientific creationism" in the United States, see Numbers, *The Creationists*. For Turkish anti-evolutionism, see Taner Edis, *An Illusion of Harmony: Science and Religion in Islam* (New York: Prometheus Books, 2010).

6. See Cornelius Hunter, *Darwin's God: Evolution and the Problem of Evil* (Ada, MI: Brazos Press, 2001).

7. Frank Sulloway, "Darwin and the Galapagos," *Biological Journal of the Linnean Society* 21, no. 1–2 (1984): 53–55.

8. Richards, "Darwin and the Descent of Woman," 58. For additional critiques of the attempts to draw a sharp boundary between Darwinism and social Darwinism, see Greene, "Darwin as a Social Evolutionist"; James Moore and Adrian Desmond, introduction to Darwin's *The Descent of Man, and Selection in Relation to Sex* (London: Penguin, 2004); and Paul, "Darwin, Social Darwinism and Eugenics."

9. Desmond and Moore, "Introduction," lv.

10. Rebecca Hardy, "David Attenborough: My Tears for My Dear Old Brother," *The Daily Mail*, February 5, 2011; John Plunkett, "Attenborough: BBC Nature Shows at Risk," *The Guardian*, January 21, 2008, https://www.theguardian.com/media/2008/jan/21/bbc.television2.

Bibliography

Anonymous. "Children." *Chambers's Edinburgh Journal* 1, no. 34 (1832): 265–66.

Anonymous. "The Department of Sanitation, Food, and Public Health." *Medical Examiner* 3 (1878): 555–56.

Anonymous. "Diphtheria and Antitoxin at the Glasgow Fever Hospital." *The Lancet* 147, no. 3798 (1896): 1668.

Anonymous. "Dr. Andrew Combe." *The Christian Reformer* 7, no. 74 (1851): 65–82.

Anonymous. "Letter of Dr. Darwin." *Gentleman's Magazine and Historical Chronicle* 104 (1808): 869.

Anonymous. Review of "The Child of the Islands. A Poem. By the Hon. Mrs. Norton." *Edinburgh Review* 82 (1845): 86–92.

Anonymous. "Review of *Vestiges of the Natural History of Creation*." *The Lancet* 1, no 2 (1845): 178.

Appel, Toby A. *The Cuvier-Geoffroy Debate: French Biology in the Decades before Darwin.* Oxford: Oxford University Press, 1987.

Aveling, Edward B. *The Religious Views of Charles Darwin.* London: Freethought Publishing Company, 1883.

Balfour, John Hutton. *Biography of the Late John Coldstream.* London: John Nisbet and Co., 1865.

Balguy, Thomas. *Divine Benevolence Asserted; And Vindicated from the Objections of Ancient and Modern Sceptics.* London: 1803. Originally published in 1781.

Barlow, Nora., ed. *The Autobiography of Charles Darwin, 1809–1822, with original omissions restored.* London: Collins, 1958.

Barrow, Mark V. *Nature's Ghosts: Confronting Extinction from the Age of Jefferson to the Age of Ecology.* Chicago: University of Chicago Press, 2009.

Bashford, Alison, and Philippa Levine, eds. *The Oxford Handbook of the History of Eugenics.* Oxford: Oxford University Press, 2010.

Basil the Great, St. *St. Basil: Ascetical Works.* Translated by M. Monica Wagner. Vol. 9 of *The Fathers of the Church: A New Translation.* Washington, DC: Catholic University of America Press, 1962.

Bates, Henry Walter. *The Naturalist on the River Amazons.* London: John Murray, 1863.

Battigelli, Anna. *Margaret Cavendish and the Exiles of the Mind.* Lexington: University Press of Kentucky, 1998.

Bernstein, Robin. *Racial Innocence: Performing American Childhood from Slavery to Civil Rights*. New York: New York University Press, 2011.

Besant, Annie. *Annie Besant: An Autobiography*, 2nd ed. London: T. Fisher Unwin, 1893. Originally published in 1885.

Bilbro, Jeffrey. "Who Are Lost and How They're Found: Redemption and Theodicy in Wheatley, Newton, and Cowper." *Early American Literature* 57 (2012): 561–89.

Blight, David W. *Frederick Douglass: Prophet of Freedom*. New York: Simon & Schuster, 2020.

Boddice, Rob. *The Science of Sympathy: Morality, Evolution, and Victorian Civilization*. Springfield: University of Illinois Press, 2016.

Bowler, Peter. "Malthus, Darwin, and the Concept of Struggle." *Journal of the History of Ideas* 37, no 4. (1976): 631–50.

Bradlaugh, Charles, and Annie Besant. "Publisher's Preface" to *Fruits of Philosophy: An Essay on the Population Question, by Charles Knowlton*, vi. Rotterdam: Van der Hoeven & Buys, 1878.

Brooke, John Hedley. *Science and Religion: Historical Perspectives*. Cambridge, UK: Cambridge University Press, 1991.

Brown, Frank Burch. "The Evolution of Darwin's Theism." *Journal of the History of Biology* 19, no. 1 (1986): 1–45.

Browne, E. Harold. *An Exposition of the Thirty-Nine Articles*, 6th ed. London: Longman, Green, Longman, Roberts & Green, 1864.

Browne, Janet. *Charles Darwin: The Power of Place*. Princeton, NJ: Princeton University Press, 2002.

———. *Charles Darwin: Voyaging*. Princeton, NJ: Princeton University Press, 1996.

Browning, Elizabeth Barrett. *The Poetical Works of Elizabeth Barrett Browning*. New York: Macmillan, 1897.

Bryan, William Jennings. *In His Image*. Chicago: Fleming H. Revell Co., 1922.

Buckland, William. *Vindiciae Geologicae: Or the Connexion of Geology with Religion Explained*. Oxford: Oxford University Press, 1820.

Burkhardt, Richard W. "The Inspiration of Lamarck's Belief in Evolution." *Journal of the History of Biology* 5, no. 2 (1972): 413–38.

———. *The Spirit of System: Lamarck and Evolutionary Biology*. Cambridge, MA: Harvard University Press, 1995.

Cameron, Lauren. "Mary Shelley's Malthusian Objections in *The Last Man*." *Nineteenth-Century Literature* 67, no. 2 (2012): 177–203.

Cantor, Geoffrey. "Science, Providence, and Progress at the Great Exhibition." *Isis* 103, no. 3 (2012): 439–59.

Cavendish, Margaret. *Observations upon Experimental Philosophy*, edited by Eugene Marshall. Indianapolis: Hackett Publishing, 2016. From the 2nd edition (London: A. Maxwell, 1668).

Chambers, Robert. *Explanations: A Sequel to "Vestiges of the Natural History of Creation."* New York: Wiley & Putnam, 1846.

———. *Vestiges of the Natural History of Creation*. London: John Churchill, 1844.

Chambers, William. *Memoir of Robert Chambers, with Autobiographic Reminiscences of William Chambers*. Edinburgh and London: W&R Chambers, 1872.

Charles, Jean Max. "The Slave Revolt That Changed the World and the Conspiracy against It: The Haitian Revolution and the Birth of Scientific Racism." *Journal of Black Studies* 51, no. 4 (2020): 275–94.

Chilton, William. "Theory of Regular Gradation." *The Movement, Anti-persecution Gazette, and Register of Progress: A Weekly Journal of Republic Politics, Anti-Theology, & Utilitarian Morals* 1 (1845): 413–14.

Clarke, Martin Lowther. *Paley: Evidences for the Man*. Toronto: University of Toronto Press, 1975.

Combe, Andrew. *Treatise on the Physiological and Moral Management of Infancy*. Philadelphia: Carey & Hart, 1840.

Combe, George. *The Life and Correspondence of Andrew Combe*. Edinburgh: MacLachlan and Stuart, 1850.

Cooke, Martin Huxley. *The Evolution of Nettie Huxley: 1825–1914*. Chichester, UK: Phillimore, 2008.

Crook, David Paul. *Darwinism, War and History: The Debate over the Biology of War from the "Origin of Species" to the First World War*. Cambridge, UK: Cambridge University Press, 1994.

Crum, Frederick S. "The Effect of Infant Mortality on the After-Lifetime of Survivors." *American Child Hygiene Association: Transactions of the Tenth Annual Meeting*, 177–94. Albany, NY: J. B. Lyon, 1920.

Cuvier, Georges. "Cuvier's Biographical Memoir of M. de Lamarck." *The Edinburgh New Philosophical Journal* 20, no. 39 (1836): 1–22.

Darwin, Charles. *1838–1839. Notebook N* [Metaphysics and Expression]. CUL-DAR126. Edited by Paul Barrett and John van Wyhe, Darwin Online. http://darwin-online .org.uk/.

———. *1837–1838. Notebook B:* [Transmutation of Species]. CUL-DARW121. Edited by John van Wyhe, Darwin Online. http://darwin-online.org.uk.

———. *The Descent of Man, and Selection in Relation to Sex*, 2nd ed. London: John Murray, 1874.

———. *Journal of Researches into the Geology and Natural History of the Various Countries Visited by H.M.S. Beagle, under the Command of Captain Fitzroy, R.N. from 1832 to 1836*, 1st ed. London: Henry Colburn, 1839.

———. *Journal of Researches into the Natural History and Geology of the Countries Visited during the Voyage of H.M.S. Beagle round the World, under the Command of Capt. Fitz Roy, R.N.*, 2nd ed. London: John Murray, 1845.

———. *On the Origin of Species by Means of Natural Selection, or the Preservation of Favoured Races in the Struggle for Life*. London: John Murray, 1859.

———. "This Is the Question Marry Not Marry" [Memorandum on marriage]. 1838. CUL-DAR210.8.2. Edited by John van Wyhe, Darwin Online. http://darwin -online.org.uk/.

Darwin, Charles, and Alfred Wallace. "On the Tendency of Species to Form Varieties; and on the Perpetuation of Varieties and Species by Natural Means of Selection." *Zoological Journal of the Linnean Society* 3, no. 9 (1858): 45–62.

Darwin, Francis, ed. *The Life and Letters of Charles Darwin*, vols. 1 and 2. London: John Murray, 1887.

Darwin, Erasmus. *The Botanic Garden*, vol. 1. London: J. Johnson, 1791.

———. *The Temple of Nature; Or, the Origin of Society: A Poem, with Philosophical Notes.* London: J. Johnson, 1803.

———. *Zoonomia, or the Laws of Organic Life*, vol. 1, 2nd ed. Boston: Thomas and Andrews, 1803.

Day, Matthew. "Godless Savages and Superstitious Dogs: Charles Darwin, Imperial Ethnography, and the Problem of Human Uniqueness." *Journal of the History of Ideas* 69, no. 1 (2008): 49–70.

de Beer, Gavin., ed. *Darwin's Notebooks on Transmutation of Species Part II. Second Notebook (February to July 1838).* London: 1960.

Delbourgo, James. "The Newtonian Slave Body: Racial Enlightenment in the Atlantic World." *Atlantic Studies* 9, no. 2 (2012): 182–207.

Desmond, Adrian. *Huxley: From Devil's Disciple to Evolution's High Priest.* Reading, MA: Addison-Wesley, 1994.

———. *The Politics of Evolution: Morphology, Medicine and Reform in Radical London.* Chicago: University of Chicago Press, 1989.

Desmond, Adrian, and James Moore. *Darwin: The Life of a Tormented Evolutionist.* New York: Norton and Company, 1991.

———. *Darwin's Sacred Cause: How a Hatred of Slavery Shaped Darwin's Views on Human Evolution.* New York: Houghton Mifflin Harcourt, 2009.

Dilbeck, D. H. *Frederick Douglass: America's Prophet.* Chapel Hill: University of North Carolina Press, 2018.

Doddridge, Philip. *Sermons to Young Persons.* London: J. Fowler, 1735.

———. *Submission to Divine Providence in the Death of Children, recommended and inforced, in a Sermon preached at Northampton, On the Death Of a very amiable and hopeful Child, about Five Years old. Published out of Compassion to mourning Parents.* London: R. Hett, 1737.

Douglass, Frederick. *The Claims of the Negro, Ethnologically Considered.* Rochester, NY: Lee, Mann & Co, 1854.

———. *The Essential Douglass: Selected Writings and Speeches.* Hackett Publishing, 2016.

Duane, Anna Mae. *Educated for Freedom: The Incredible Story of Two Fugitive Schoolboys Who Grew Up to Change a Nation.* New York: New York University Press, 2020.

Durant, John R. "Scientific Naturalism and Social Reform in the Thought of Alfred Russel Wallace." *British Journal for the History of Science* 12, no. 1 (1979): 31–58.

Edis, Taner. *An Illusion of Harmony: Science and Religion in Islam.* New York: Prometheus Books, 2010.

Fara, Patricia. *Erasmus Darwin: Sex, Science, and Serendipity.* Oxford: Oxford University Press, 2012.

Fichman, Martin, and Jennifer E. Keelan. "Resister's Logic: The Anti-Vaccination Arguments of Alfred Russel Wallace and Their Role in the Debates over Compulsory Vaccination in England, 1870–1907." *Studies in History and Philosophy of Science Part C: Studies in History and Philosophy of Biological and Biomedical Sciences* 38, no. 3 (2007): 585–607.

Fitzroy, Robert. *Narrative of the Surveying Voyages of His Majesty's Ships Adventure and Beagle, Between the Years 1826–36, Describing Their Examination of the Southern Shores of South America, and the Beagle Circumnavigation of the Globe: In Three Volumes. Proceedings of the second expedition, 1831–1836, under the command of Captain Robert Fitz-Roy.* Volume 2. London: Henry Colburn, 1839.

Franklin, Benjamin. "The Death of Infants." *Pennsylvania Gazette*, 1734.

Fuller, Randall. *The Book That Changed America: How Darwin's Theory of Evolution Ignited a Nation.* New York: Penguin, 2018.

Gale, Barry G. "Darwin and the Concept of a Struggle for Existence: A Study in the Extra-Scientific Origins of Scientific Ideas." *Isis* 63, no. 3 (1972): 321–44.

Galton, Francis. "Eugenics: Its Definition, Scope, and Aims." *The American Journal of Sociology* 10, no. 1 (1904): 1–25.

———. *Memories of My Life.* Yorkshire: Methuen, 1908.

Gandhi, Mahatma. *A Guide to Health.* Madras: S. Ganesan, 1921.

Garfinkle, Norton. "Science and Religion in England, 1790–1800: The Critical Response to the Work of Erasmus Darwin." *Journal of the History of Ideas* 16, no. 3 (1955): 376–88.

Gillespie, Neil C. "Divine Design and the Industrial Revolution: William Paley's Abortive Reform of Natural Theology." *Isis* 81, no. 2 (1990): 214–29.

———. "Natural History, Natural Theology, and Social Order: John Ray and the 'Newtonian Ideology.'" *Journal of the History of Biology* 20 (1987): 1–49.

Grant, Charles William. *Our Blood Relations, or the Darwinian Theory.* London: Simpkin, Marshall & Co., 1872.

Graves, Charles Larcom. *The Life & Letters of Sir George Grove.* London: Macmillan and Co., 1903.

Gray, Asa. *Darwiniana: Essays and Reviews Pertaining to Darwinism.* New York: D. Appleton and Company, 1877.

———. *Natural Selection not Inconsistent with Natural Theology: A Free Examination of Darwin's Treatise on the Origin of Species, and of its American Reviewers.* Reprinted from *The Atlantic Monthly* (July, August, and October, 1860). London: Trübner and Co., 1861.

———. "Review of *Vestiges of the Natural History of Creation*, by Robert Chambers." *The North American Review* 60, no. 127 (1845): 426–78.

Greene, John C. "Darwin as a Social Evolutionist." *Journal of the History of Biology* 10, no. 1 (1977): 1–27.

Gulick, John. *Evolutionist and Missionary, John Thomas Gulick, Portrayed through Documents and Discussions.* Chicago: University of Chicago Press, 1932.

Haeckel, Ernst. *The Evolution of Man: A Popular Study.* New York: G. P. Putnam's Sons, 1910.

Hague, J. D. "A Reminiscence of Mr. Darwin." *Harper's New Monthly Magazine* 69, issue 413 (1884): 759–63.

Hale, Piers J. *Political Descent: Malthus, Mutualism, and the Politics of Evolution in Victorian England.* Chicago: University of Chicago Press, 2014.

Hall, Marshall. *The Two-Fold Slavery of the United States.* London: Adam Scott, 1854.

Haller, John S. "The Physician versus the Negro: Medical and Anthropological Concepts of Race in the Late Nineteenth Century." *Bulletin of the History of Medicine* 44, no. 2 (1970): 154–67.

Hammonds, Evelynn. *Childhood's Deadly Scourge: The Campaign to Control Diphtheria in New York City, 1880–1930.* Baltimore, MD: Johns Hopkins University Press, 2002.

Hardy, Rebecca. "David Attenborough: My Tears for My Dear Old Brother." *The Daily Mail*, February 5, 2011.

Harrison, Peter. *The Fall of Man and the Foundations of Science.* Cambridge, UK: Cambridge University Press, 2007.

Henslow, John Stevens. *A Sermon on the First and Second Resurrection.* Cambridge, UK: James Hodson, 1829.

Hiroa, Te Rangi (P. H. Buck). "The Passing of the Maori." *Transactions and Proceedings of the New Zealand Institute* 55 (1924): 362–75.

Hodgson, Dennis. "Malthus' Essay on Population and the American Debate over Slavery." *Comparative Studies in Society and History* 51, no. 4 (2009): 742–70.

Hofstadter, Richard. *Social Darwinism in American Thought: 1860–1915.* Philadelphia: University of Pennsylvania Press, 1944.

Holland, Frederic May. *Frederick Douglass: The Colored Orator.* New York: Funk and Wagnalls Company, 1895.

Holyoake, George Jacob. *The History of the Last Trial by Jury for Atheism in England.* London: James Watson, 1850.

———. *Paley Refuted in His Own Words.* London: Hetherington, 1847.

Hunter, Cornelius. *Darwin's God: Evolution and the Problem of Evil.* Ava, MI: Brazos Press, 2001.

Hunter, Michael. *Science and the Shape of Orthodoxy: Intellectual Change in Late Seventeenth-Century Britain.* Woodbridge, UK: Boydell Press, 1995.

Hutchins, Zachary Mcleod. "Building Bensalem at Massachusetts Bay: Francis Bacon and the Wisdom of Eden in Early Modern New England." *The New England Quarterly* 83 (2010): 577–606.

Huxley, Leonard, ed. *Life and Letters of Sir Joseph Dalton Hooker.* London: John Murray, 1918.

———, ed. *Life and Letters of Thomas Henry Huxley*, vol. 1. London: Macmillan and Co., 1900.

———, ed. *Life and Letters of Thomas Henry Huxley*, vol. 3. London: Macmillan and Co., 1913.

Huxley, Thomas Henry. *Evolution and Ethics, and Other Essays.* New York: D. Appleton and Company, 1894.

———. "Report on Spiritualism." *Daily News*, October 17, 1871.

————. "Review VII. *Vestiges of the Natural History of Creation*, Tenth Edition, London, 1853." *The British and Foreign Medico-Chirurgical Review* 13, no. 13 (1854): 425–29.

————. *Science and Education: Essays*. New York: D. Appleton and Company, 1898.

James, Patricia. *Population Malthus: His Life and Times*. London: Routledge, 2013.

Jefferson, Thomas. *Notes on the State of Virginia*. Boston: Lilly and Wait, 1832.

Jensen, J. Vernon. "Return to the Wilberforce–Huxley Debate." *The British Journal for the History of Science* 21, no. 2 (1988): 161–79.

Jenyns, Leonard. *Memoir of the Rev. John Stevens Henslow*. London: John Van Voorst, 1862.

Jones, David S. *Rationalizing Epidemics: Meanings and Uses of American Indian Mortality since 1600*. Cambridge, MA: Harvard University Press, 2004.

Kass, Amalie M., and Edward Harold Kass. *Perfecting the World: The Life and Times of Dr. Thomas Hodgkin 1798–1866*. Boston: Harcourt Brace Jovanovich, 2008.

Keynes, Randal D., ed. *Charles Darwin's* Beagle *Diary*. Cambridge, UK: Cambridge University Press, 1988.

————. *Darwin, His Daughter, and Human Evolution*. New York: Penguin, 2002.

Kidd, John. *On the Adaptation of External Nature to the Physical Condition of Man, Principally with Reference to the Supply of His Wants, and the Exercise of His Intellectual Faculties*. Philadelphia: Carey, Lea & Blanchard, 1836.

King-Hele, Desmond, ed. *The Collected Letters of Erasmus Darwin*. Cambridge, UK: Cambridge University Press, 2007.

Kirby, William. *On the Power, Wisdom and Goodness of God as Manifested in the Creation of Animals and their History, Habits, and Instincts*, vol. 1. London: William Pickering, 1835.

Kohn, David. "Darwin's Ambiguity: The Secularization of Biological Meaning." *The British Journal for the History of Science* 22, no. 2 (1989): 215–39.

Kottler, Malcolm Jay. "Alfred Russel Wallace, the Origin of Man, and Spiritualism." *Isis* 65, no. 2 (1974): 145–92.

Laidlaw, Zoë. "Heathens, Slaves and Aborigines: Thomas Hodgkin's Critique of Missions and Anti-Slavery." *History Workshop Journal* 64, no. 1 (2007): 133–61.

Larson, Edward J. *Summer for the Gods: The Scopes Trial and America's Continuing Debate over Science and Religion*. New York: Basic Books, 1997.

Le Mahieu, Dan Lloyd. *The Mind of William Paley: A Philosopher and His Age*. Lincoln: University of Nebraska Press, 1976.

Lehmann, R. C. "Memories of Half a Century." *Chambers's Journal* 80 (1903): 49–52.

Lightman, Bernard. "Huxley and the Devonshire Commission." In *Victorian Scientific Naturalism*, edited by Gowan Dawson and Bernard Lightman, 101–30. Chicago: University of Chicago Press, 2014.

————. *The Origins of Agnosticism: Victorian Unbelief and the Limits of Knowledge*. Baltimore: Johns Hopkins University Press, 1987.

Lindberg, David. "Galileo, the Church, and the Cosmos." In *When Science and Christianity Meet*, edited by David Lindberg and Ronald L. Numbers, 33–66. Chicago: University of Chicago Press, 2003.

Lindberg, David C., and Ronald L. Numbers. "Beyond War and Peace: A Reappraisal of the Encounter between Christianity and Science." *Church History* 55, no. 3 (1986): 338–54.

Litchfield, Henrietta E. *Emma Darwin: Wife of Charles Darwin*, vol. 2. Cambridge, UK: Cambridge University Press, 1904.

Livingstone, David N. *Darwin's Forgotten Defenders: The Encounter between Evangelical Theology and Evolutionary Thought*. Vancouver: Regent College Publishing, 1997.

Lowndes, Thomas. *Tracts, Political and Miscellaneous, in Prose and Verse*. London: 1827.

Lucas, John Randolph. "Wilberforce and Huxley: A Legendary Encounter." *The Historical Journal* 22, no. 2 (1979): 313–30.

Lyell, Charles. *The Geological Evidences of the Antiquity of Man, with Remarks on Theories of the Origin of Species by Variation*. London: John Murray, 1863.

———. *Principles of Geology: Being an Inquiry How Far the Former Changes of the Earth's Surface are Referable to Causes now in Operation*, vol. 2. London: John Murray, 1833.

Malthus, Thomas. *An Essay on the Principle of Population*, vol. 2. London: John Murray, 1826.

———. *First Essay on Population 1798*, with notes by James Bonar. London: MacMillan and Co., 1926.

Mandelbaum, Maurice. "Darwin's Religious Views." *Journal of the History of Ideas* 19, no. 3 (1958): 363–78.

Marchant, James, ed. *Alfred Russel Wallace: Letters and Reminiscences*, vols. 1 and 2. London: Cassell and Company, 1916.

Marshall, Ashley. "Erasmus Darwin *contra* David Hume." *British Journal for Eighteenth-Century Studies* 30, no. 1 (2007): 89–111.

Marshall, Julian. *The Life and Letters of Mary Wollstonecraft Shelley*, vol. 1. London: Richard Bentley & Son, 1889.

Martineau, Harriet. *Harriet Martineau's Autobiography*. London: Smith, Elder & Co., 1877.

Mayr, Ernst. *The Growth of Biological Thought: Diversity, Evolution, and Inheritance*. Cambridge, MA: Harvard University Press, 1982.

Meadley, G. W. "Memoirs of William Paley, D.D." In *The Works of William Paley*, vol. 1. Boston: J. Graham, 1810.

Miller, Hugh. *Foot-Prints of the Creator: or, The Asterolepis of Stromness*. London: Johnstone and Hunter, 1849.

Millward, Robert, and Frances Bell. "Infant Mortality in Victorian Britain: The Mother as Medium." *Economic History Review* 54, no. 4 (2001): 699–733.

Moore, James R. *The Darwin Legend*. Grand Rapids, MI: Baker Books, 1994.

———. "Of Love and Death: Why Darwin 'Gave Up Christianity.'" In *History, Humanity, and Evolution: Essays for John C. Greene*, edited by James R. Moore, 195–229. Cambridge, UK: Cambridge University Press, 1989.

———. *The Post-Darwinian Controversies: A Study of the Protestant Struggle to Come to Terms with Darwin in Great Britain and America, 1870–1900*. Cambridge, UK: Cambridge University Press, 1981.

———. "Theodicy and Society: The Crisis of the Intelligentsia." In *Victorian Faith in Crisis*, edited by Richard J. Helmstadter and Bernard Lightman, 153–86. London: Palgrave Macmillan, 1990.

Moore, James, and Adrian Desmond. Introduction to Darwin's *The Descent of Man, and Selection in Relation to Sex*, xi–lxiv. London: Penguin, 2004.

Morris, Caspar. *Lectures on Scarlet Fever*. Philadelphia: Lindsay & Blakiston, 1851.

Morrison, Larry R. "The Religious Defense of American Slavery before 1830." *The Journal of Religious Thought* 37, no. 2 (1980): 16–29.

Murphy, Howard R. "The Ethical Revolt against Christian Orthodoxy in Early Victorian England." *The American Historical Review* 60, no. 4 (1955): 800–817.

Nadler, Steven M. *Spinoza: A Life*. Cambridge, UK: Cambridge University Press, 2001.

Newsholme, Arthur. *Fifty Years in Public Health: A Personal Narrative with Comments*. London: George Allen & Unwin, 1935.

Newton, Hannah. "The Dying Child in Seventeenth-Century England." *Pediatrics* 136, no. 2 (August 2015): 218–20.

———. *The Sick Child in Early Modern England, 1580–1720*. Oxford: Oxford University Press, 2012.

Noble, David. *The Religion of Technology: The Divinity of Man and the Spirit of Invention*. New York: Alfred K. Knopf, 1997.

———. *A World without Women: The Christian Clerical Culture of Western Science*. New York: Alfred K. Knopf, 1992.

Nott, Josiah C. "The Mulatto a Hybrid—Probable Extermination of the two races if Whites and Blacks are allowed to Marry." *The American Journal of the Medical Sciences* 5 (1843): 252–56.

Nott, Josiah C., and George R. Gliddon. *Types of Mankind*. London: Trübner & Co., 1854.

Numbers, Ronald L. *The Creationists: From Scientific Creationism to Intelligent Design*. Cambridge, MA: Harvard University Press, 2006.

———. *Science and Christianity in Pulpit and Pew*. Oxford: Oxford University Press, 2007.

———, ed. *Galileo Goes to Jail and Other Myths about Science and Religion*. Cambridge, MA: Harvard University Press, 2010.

Nye, Robert A. "The Rise and Fall of the Eugenics Empire: Recent Perspectives on the Impact of Biomedical Thought in Modern Society." *The Historical Journal* 36, no. 3 (1993): 687–70.

Orme, Nicholas. *Medieval Children*. New Haven, CT: Yale University Press, 2001.

Paley, Edmund. *The Works of William Paley, D.D. with Additional Sermons, etc. etc. and a Corrected Account of the Life and Writings of the Author*, vol. 7. London: C. and J. Rivington, etc., 1825.

Paley, William. *Moral and Political Philosophy*. New York: S. King, 1824.

———. *Natural Theology: or, Evidences of the Existence and Attributes of the Deity, Collected from the Appearances of Nature*. London: R. Faulder, 1802.

———. *The Works of William Paley, D.D. Archdeacon of Carlisle. With a Life of the Author*, vol. 3. London: Thomas Tegg, 1825.

———. *The Works of William Paley, D.D. Archdeacon of Carlisle*. Edinburgh: Peter Brown and Thomas Nelson, 1829.

Parker, T. Jeffery. "Professor Huxley: From the Point of View of a Disciple." *Natural Science: A Monthly Review of Scientific Progress* 8 (1896): 161–67.

Paul, Diane B. *Controlling Human Heredity: 1865 to the Present*. New York: Prometheus Books, 1995.

———. "Darwin, Social Darwinism and Eugenics." In *The Cambridge Companion to Darwin*, edited by Jonathan Hodge and Gregory Radick, 214–39. Cambridge, UK: Cambridge University Press, 2006.

———. "Eugenics and the Left." *Journal of the History of Ideas* 45, no. 4 (1984): 567–90.

Peabody, William B. O. *The Loss of Children: A Sermon, delivered January 22, 1832*. S. Bowler, 1832.

Pearson, John. *An Exposition of the Creed*. Oxford: Clarendon Press, 1797.

Picciotto, Joanna. "Reforming the Garden: The Experimentalist Eden and 'Paradise Lost.'" *English Literary History* 72, no. 1 (April 1, 2005): 23–78.

Plunkett, John. "Attenborough: BBC Nature Shows at Risk." *The Guardian*, January 21, 2008. https://www.theguardian.com/media/2008/jan/21/bbc.television2.

Porter, Roy. *English Society in the Eighteenth Century*. New York: Penguin, 1982.

———. *The Greatest Benefit to Mankind: A Medical History of Humanity from Antiquity to the Present*. London: HarperCollins, 1997.

Powell, Baden. "On the Study of the Evidences of Christianity." In *Essays and Reviews*, 94–144, 2nd ed. London: John W. Parker & Son, 1860.

Priestley, Joseph. *The Theological and Miscellaneous Works of Joseph Priestley*, vol. 18. G. Smallfield, 1787.

Prince, Mary. *The History of Mary Prince, a West Indian Slave. Related by Herself*, 3rd ed. London: F. Westley and A. H. Davis, 1831.

Raby, Peter. *Alfred Russel Wallace: A Life*. London: Chatto and Windus, 2001.

Raven, Charles. *John Ray, Naturalist: His Life and Works*. Cambridge, UK: Cambridge University Press, 1942.

Ray, John. *The Correspondence of John Ray: Consisting of Selections from the Philosophical Letters Published by Dr. Derham, and Original Letters of John Ray, in the Collection of the British Museum*. London: Ray Society, 1848.

———. *Three Physico-Theological Discourses*, 3rd ed. London: William Innys, 1713.

———. *The Wisdom of God Manifested in the Works of the Creation*. London: Samuel Smith, 1691.

Richards, Evelleen. "Darwin and the Descent of Woman." In *The Wider Domain of Evolutionary Thought*, edited by David Oldroyd and Ian Langham, 57–111. Dordrecht: Springer, 1983.

———. *Darwin and the Making of Sexual Selection*. Chicago: University of Chicago Press, 2017.

Richards, Robert J. *The Tragic Sense of Life: Ernst Haeckel and the Struggle over Evolutionary Thought*. Chicago: University of Chicago Press, 2008.

———. *Was Hitler a Darwinian? Disputed Questions in the History of Evolutionary Theory*. Chicago: University of Chicago Press, 2013.

Ristaino, Jean Beagle, and Donald H. Pfister. "'What a Painfully Interesting Subject': Charles Darwin's Studies of Potato Late Blight." *BioScience* 66, no. 12 (2016): 1035–45.

Rosen, Christine. *Preaching Eugenics: Religious Leaders and the American Eugenics Movement*. Oxford: Oxford University Press, 2004.

Ruse, Michael. *The Darwinian Revolution: Science Red in Tooth and Claw*. Chicago: University of Chicago Press, 1979.

———. *The Problem of War: Darwinism, Christianity, and the Battle to Understand Human Conflict*. Oxford: Oxford University Press, 2019.

Saleeby, Caleb Williams. *Parenthood and Race Culture: An Outline of Eugenics*. New York: Moffat, Yard and Company, 1915. First printing 1909.

Sarasohn, Lisa T. *The Natural Philosophy of Margaret Cavendish: Reason and Fancy during the Scientific Revolution*. Baltimore, MD: Johns Hopkins University Press, 2010.

Schiebinger, Londa. *The Mind Has No Sex? Women in the Origins of Modern Science*. Cambridge, MA: Harvard University Press, 1991.

Scott, Elizabeth. *Specimens of British Poetry: Chiefly Selected from Authors of High Celebrity, and Interspersed with Original Writings*. Edinburgh: James Ballantyne and Co., 1823.

Secord, James. "Behind the Veil: Robert Chambers and Vestiges." In *History, Humanity and Evolution: Essays for John C. Greene*, edited by James Moore, 165–94. Cambridge: Cambridge University Press, 1989.

———. *A Victorian Sensation: The Extraordinary Publication, Reception, and Secret Authorship of* Vestiges of the Natural History of Creation. Chicago: University of Chicago Press, 2000.

Sedgwick, Adam. "President's Address." *Proceedings of the Geological Society of London* 1, no. 15 (1830): 187–212.

——— (published anonymously). "Review of *Vestiges of the Natural History of Creation*." *The Edinburgh Review* 82 (1845): 1–85.

Seth, Suman. "Darwin and the Ethnologists: Liberal Racialism and the Geological Analogy." *Historical Studies in the Natural Sciences* 46, no. 4 (2016): 490–527.

Shapin, Steven, and Simon Schaffer. *Leviathan and the Air-Pump: Hobbes, Boyle and the Experimental Life*. Princeton, NJ: Princeton University Press, 1989.

Shelley, Mary. *Frankenstein, or the Modern Prometheus*. London: George Routledge and Sons Ltd., 1891.

Shelley, Percy Bysshe. *A Refutation of Deism: In a Dialogue*. London: Progressive Publishing Company, 1890. Originally published 1814.

Sivasundaram, Sujit. *Nature and the Godly Empire: Science and Evangelical Mission in the Pacific, 1795–1850*. Cambridge, UK: Cambridge University Press, 2005.

Smith, James McCune. "Colored Orphan's Asylum: Physicians Report." *The Colored American*, January 26, 1839.

Southwood Smith, T. *Illustrations of the Divine Government*, 1st American ed. Boston: Benj. B Mussey, 1831.

Sreenivas, Mytheli. "Birth Control in the Shadow of Empire: The Trials of Annie Besant, 1877–1878." *Feminist Studies* 41, no. 3 (2015): 509–37.

Stanley, Matthew. *Huxley's Church and Maxwell's Demon: From Theistic Science to Naturalistic Science*. Chicago: University of Chicago Press, 2015.

Stauffer, John, ed. *The Works of James McCune Smith: Black Intellectual and Abolitionist*. Oxford: Oxford University Press, 2006.

Steckel, Richard H. "A Dreadful Childhood: The Excess Mortality of American Slaves." *Social Science History* 10, no. 4 (1986): 427–65.

Stephens, Marguerita. *White without Soap: Philanthropy, Caste and Exclusion in Colonial Victoria 1835–1888: A Political Economy of Race*. University of Manitoba Custom Book Centre, 2010.

Strauss, David Friedrich. *The Old Faith and the New: A Confession*, 2nd ed. London: Asher & Co., 1873.

Sulloway, Frank J. "Darwin and His Finches: The Evolution of a Legend." *Journal of the History of Biology* 15, no. 1 (1982): 1–53.

———. "Darwin and the Galapagos." *Biological Journal of the Linnean Society* 21, no. 1–2 (1984): 29–59.

Tennyson, Alfred. *In Memoriam*, 5th ed. London: Edward Moxon, 1851.

Thornton, Alice. *The Autobiography of Mrs. Alice Thornton, of East Newton Co. York*. Durham, UK: Andrews and Co., 1875.

Tobin, Jeffrey. "Savages, the Poor and the Discourse of Hawaiian Infanticide." *The Journal of the Polynesian Society* 106, no. 1 (1997): 65–92.

Tomes, Nancy. *The Gospel of Germs: Men, Women, and the Microbe in American Life*. Cambridge, MA: Harvard University Press, 1999.

Tourgée, Albion Winegar. *An Appeal to Caesar*. New York: Fords, Howard and Hulbert, 1884.

Turner, James. *Without God, without Creed: The Origins of Unbelief in America*. Baltimore, MD: Johns Hopkins University Press, 1985.

Wallace, Alfred Russel. *A Defence of Modern Spiritualism*. Boston: Colby and Rich, 1874.

———. *On Miracles and Modern Spiritualism*. London: James Burns, 1875.

———. *My Life: A Record of Events and Opinions*, vol. 1. New York: Dodd, Mead and Company, 1905.

———. *A Narrative of Travels on the Amazon and Rio Negro, with an account of the native tribes, and observations of the climate, geology, and natural history of the Amazon valley*. London: Reeve and Co., 1853.

———. "The Origin of Human Races and the Antiquity of Man Deduced From the Theory of 'Natural Selection.'" *Journal of the Anthropological Society of London* 2 (1864): clviii–clxxxvii.

———. "Sir Charles Lyell on Geological Climates and the Origin of Species." *Quarterly Review* 126, no. 252 (1869): 359–94.

———. *The Wonderful Century: Its Successes and Its Failures*. London: Swann Sonnenschein, 1898.

Waterhouse, Benjamin. *An Essay Concerning Tussis Convulsiva, or, Whooping Cough. With Observations on the Diseases of Children*. Boston: Munroe and Francis, 1822.

Weber, Thomas P. "Alfred Russel Wallace and the Antivaccination Movement in Victorian England." *Emerging Infectious Disease* 16, no. 4 (2010): 664–68.

Weikart, Richard. *From Darwin to Hitler: Evolutionary Ethics, Eugenics and Racism in Germany*. Dordrecht: Springer, 2016.

Wellman, Kathleen Anne. *La Mettrie: Medicine, Philosophy, and Enlightenment*. Durham, NC, and London: Duke University Press, 1992.

Westfall, Richard. *Science and Religion in Seventeenth-Century England*. New Haven, CT: Yale University Press, 1958.

Wheatley, Phillis. *Memoir and Poems of Phillis Wheatley, A Native African and a Slave*. Boston: Geo W. Light, 1834.

Wilberforce, Samuel. "Review of *On the Origin of Species*." *Quarterly Review* 108 (1860): 118–38.

Williams, A. N. "Physician, Poet, Polymath and Prophet: Erasmus Darwin a New Window Eighteenth-Century Child Health Care." *British Society for the History of Paediatrics and Child Health* 96, suppl. 1 (2011): A43.

Williams, Heather Andrea. *Help Me to Find My People: The African American Search for Family Lost in Slavery*. Chapel Hill: University of North Carolina Press, 2012.

Wilson, Leonard G., ed. *Sir Charles Lyell's Scientific Journals on the Species Question*. New Haven, CT: Yale University Press, 1970.

Van Wyhe, John, and Mark J. Pallen. "The 'Annie Hypothesis': Did the Death of His Daughter Cause Darwin to 'Give Up Christianity'?" *Centaurus* 54, no. 2 (2012): 105–23.

Yannielli, Joseph L. "A Yahgan for the Killing: Murder, Memory and Charles Darwin." *The British Journal for the History of Science* 46, no. 3 (2013): 415–43.

Young, Robert. "Malthus and the Evolutionists: The Common Context of Biological and Social Theory." *Past and Present* 43 (1969): 109–45.

Index

abolition, 45–46, 72, 97, 111, 222, 224–25; Asa Gray and, 173; Charles Darwin and, 90, 95–96, 99–100, 204; Darwin-Wedgwood family and, 64, 96–96; Erasmus Darwin and, 44–46; Frederick Douglass and, 219; Harriet Martineau and, 127; Phillis Wheatley and, 13–15; Thomas Henry Huxley and, 222. *See also* slavery

Aborigines Protection Society, 113

agnosticism, 156, 188–89, 191–92

Albert, Prince, 142

anatomy, 7–8, 16, 18–20, 22, 59, 60–62, 65, 76

anesthesia, 59

Anglican Church: Alfred Russel Wallace and, 155, 197; Annie Besant and, 228–29; Charles Darwin and, 76–77, 112, 130–31; doctrines of, viii–ix; 35–36, 77, 150–51; Emma Darwin and, 132–33; history of, 4; missions of, 102, 112; natural theology and, 59, 80–81, 120, 126; Thomas Henry and Nettie

Huxley and, 180–81, 183–184, 186, 242

Anthropological Society of London, 198–99

Anti-Jacobin Review, 53

antitoxin, 241–42

appendix, vermiform, 201–2

Aries, Philip, viii

atheism, 2; Adam Sedgwick and, 87; Annie Besant and, 227–30, 232, 234–35; association with evolution, 126, 137; Charles Darwin and, 136, 147–48, 150; Erasmus Darwin and, 32–33; mechanical philosophy and, 5–7; Percy Shelley and, 57–58; Phillis Wheatley and, 14; Thomas Henry Huxley and, 188

Atkinson, Henry George, 148

Attenborough, David, 254

Augustine, St., 17

Autobiography of Charles Darwin (Darwin), 147–48, 150, 151, 168–69

Aveling, Edward, 147